Praise for
A Theory of Everything (That Matters)

This brief book provides an answer to those who wonder, What is the value of religion in a modern scientific world? McGrath gives an accessible introduction to Einstein's science, then reflects on how the great physicist synthesized politics, ethics, science, and religion in his view of the world. Einstein felt that, while science can help us achieve our moral goals, science alone cannot create those moral goals or the means to instill them in society. McGrath shows that we need both religion and science to best engage the mysteries of the world and everything that matters.

DEBORAH HAARSMA
Astrophysicist and president of BioLogos

Albert Einstein was, without doubt, the most iconic scientist of the last century. In this lucid little book, Alister McGrath provides an accessible introduction to Einstein's great scientific discoveries, as well as a careful analysis of his views on the relationship between science and religion. Einstein was a nuanced thinker on the big questions of life, and who better than McGrath to guide us on an exploration of this aspect of Einstein's legacy? I recommend this book to anyone who wants a fuller picture of Albert Einstein's life and thought.

ARD LOUIS

Professor of theoretical physics, University of Oxford

a theory of everything
(that matters)

ALISTER
McGRATH

a theory
of
everything
(that matters)

A BRIEF GUIDE TO EINSTEIN,
RELATIVITY & HIS SURPRISING
THOUGHTS ON GOD

TYNDALE
MOMENTUM®

*The nonfiction imprint of
Tyndale House Publishers, Inc.*

Visit Tyndale online at www.tyndale.com.

Visit Tyndale Momentum online at www.tyndalemomentum.com.

TYNDALE, Tyndale's quill logo, and *Tyndale Momentum* are registered trademarks of Tyndale House Publishers, Inc. The Tyndale Momentum logo is a trademark of Tyndale House Publishers, Inc. Tyndale Momentum is the nonfiction imprint of Tyndale House Publishers, Inc., Carol Stream, Illinois.

A Theory of Everything (That Matters): A Brief Guide to Einstein, Relativity, and His Surprising Thoughts on God

Designed by Ron C. Kaufmann

Edited by Jonathan Schindler

For information about special discounts for bulk purchases, please contact Tyndale House Publishers at csresponse@tyndale.com, or call 1-800-323-9400.

ISBN 978-1-4964-3807-2

Printed in the United States of America

25	24	23	22	21	20	19
7	6	5	4	3	2	1

contents

ALBERT EINSTEIN:
THE WORLD'S FAVORITE
GENIUS

ALBERT EINSTEIN REMAINS the world's favorite genius, propelled to fame by popular adulation of his revolutionary scientific theories about space and time. Today, a century after the confirmation of his theory of general relativity in November 1919, Einstein remains a cult figure. He has appeared on the cover of *Time* magazine no fewer than six times and was lionized as its Person of the Century in 1999. His equation $E = mc^2$ has become the best-known scientific formula of all time and has regularly—along with Einstein's trademark hairstyle—found its way onto T-shirts and billboards.

Photographers loved Einstein. One of the best-known photos of him is Arthur Sasse's shot of Einstein sticking out his tongue. This iconic photograph was taken right at the end of his birthday party in 1951 at Princeton, as a weary Einstein entered his chauffeured automobile to be driven home. Sasse, who had been covering the event, ran up to the open door and asked Einstein for one final shot. Einstein turned toward him and stuck out his tongue just as Sasse's flashbulb went off. Einstein liked the resulting photo so much that he used it for greeting cards he sent to his friends.

Einstein's ideas have changed the way we think and live. Without realizing it, we depend on his theory of relativity when using a Global Positioning System (GPS). The light and warmth of the sun are the direct result of the conversion of mass to energy—the process Einstein first recognized in 1905 and expressed in his equation $E = mc^2$. This same principle lies behind nuclear power generators—and atomic bombs. Einstein triggered America's race to build the atomic bomb in 1939 with a letter to President Franklin D. Roosevelt, warning him that Nazi Germany would get there first unless the United States committed itself to developing the necessary technology. (A copy of this typewritten letter was sold at Christie's in New York for $2.1 million in 2008.)

The immense esteem in which Einstein was held by the academic and popular-science community meant that when he talked about larger questions, people were prepared to listen. When his theory of general relativity was triumphantly confirmed in November 1919, he became a media sensation. His 1921 tour of America was front-page news.

Yet Einstein did not simply speak about science. He opened up grander issues of human value and meaning—what the philosopher Karl Popper later called "ultimate questions." People listened to Einstein with attentiveness and respect. He became a celebrity genius, an intellect of colossal status, who managed to achieve iconic cultural status without dumbing down what he said.

Einstein doesn't fit the stuffy stereotype of a scientific genius. On a visit to California, he struck up a surprising friendship with the movie star Charlie Chaplin. Chaplin invited Einstein to attend the premiere of his 1931 movie *City Lights*. The huge crowd went wild as Einstein and Chaplin arrived together. According to a popular legend, Chaplin told Einstein, "They're cheering you because nobody understands you, and me because everybody understands me." Although Einstein was awarded the 1921 Nobel Prize in Physics, his greatest achievement was arguably to become admired, even adored, by the wider public. Many sensed that although Einstein was difficult to understand, he had grasped something profound about our universe that others had failed to find. He was worth listening to—even if doing so was difficult and demanding.

This short book sets out to explain in simple and accessible terms Einstein's revolutionary scientific ideas, which still shape our world today, and to explore their significance. Nobody thinks a scientific genius is infallible. Still, Einstein's status means he is profoundly worth listening to, especially when thinking about how we make sense of our universe.

Yet this book goes further—as, I believe, Einstein would wish us to do. It takes seriously Einstein's fascination with a "big picture" of our world and ourselves. It considers how Einstein personally wove together science, ethics, and religious faith to yield a richer account of reality—if you like, a theory of everything *that matters*. So how did he do this? What were the outcomes? What remains valuable for us today? Lots of people have written about Einstein's wider vision of life, stimulating and informing our discussion. So why not make Einstein our dialogue partner? After all, he was a genius.

Finally, a word of caution. Many sayings attributed to Einstein have no connection with him whatsoever. Here's one that is regularly attributed to Einstein, which I first came across on a T-shirt at an American university: "Not everything that can be counted counts, and not everything that counts can be counted." It's a great idea—but it's not Einstein's. Throughout this book, I have tried to ensure reliable citation of Einstein's works. Here's an authentic quote that sets up the agenda of this book: "Science can only ascertain what *is*, but not what *should be*."[1] Einstein here invites us to explore how to hold science and moral thinking together and sets out his own way of doing this. Science, ethics, and religion are quite different undertakings, playing distinct roles in our lives and based on different thought patterns. How then can we weave their unique perspectives together into a coherent whole? It's a genuine issue, and Einstein helps us to think about this.

Although this book explores Einstein's scientific ideas, its real focus is how he attempted to develop a coherent view of the world—a "grand theory of everything" that embraces both our understanding of how the world *functions* and the deeper question of what it *means*. Einstein wasn't just a great scientist; he was a reflective human being who realized the importance of holding together our key ideas and beliefs. His reflections on how to develop such a "big picture" of our world and ourselves might help us move beyond the fragmentation of ideas and values that has become such a core feature of Western culture.

But enough has been said by way of introduction. It's time to engage with Einstein himself.

chapter 1

APPROACHING EINSTEIN: THE WONDER OF NATURE

SCIENCE RARELY MAKES the headlines in British news-papers. But in 1919, a year after the end of the Great War, that changed decisively. On Friday, November 7, the London *Times* printed a headline above a report on a dramatic new development: "Revolution in Science. New Theory of the Universe. Newtonian Ideas Overthrown." Like much scientific journalism, this headline was sensationalist,[1] suggesting that, just as in the then-recent political and social revolutions in China in 1911 and Russia in 1917, an old order had been swept away. Sir Isaac Newton—widely regarded as the greatest British scientist—had been dethroned, his ideas now discredited, lying in tatters. And who was the cause of this revolution in science? An obscure German physicist, hitherto unknown to the readership of the *Times*—Albert Einstein.

The *Times* headline propelled Einstein to international celebrity. It was an extraordinary moment. Britain and Germany had only just emerged from the most destructive and traumatic total war yet known, which had created distrust and hatred between the two nations on an unprecedented scale. Yet almost exactly one year after the end of the First World War, Britain's scientific elite had embraced Einstein, a German national and former enemy, in the common human search for an understanding of our universe. It seemed to be a symbol of hope in the bleak postwar era. Might international scientific cooperation hold the key to new understandings of our world and ourselves? Einstein found himself propelled into the limelight. A disillusioned and restless postwar generation seized on him as someone who could finally make sense of our perplexing world and our place within it.

By the early 1920s, Einstein had become a cult figure, an international icon of genius, helped to no small extent by the award of the 1921 Nobel Prize in Physics—and perhaps also by his distinctive appearance. Einstein made fuzzy hair a hallmark of intelligence. (At a Hollywood dinner party in the winter of 1931–32, the movie actress Marion Davies ruffled Einstein's notoriously unruly hair and quipped, "Why don't you get your hair cut?") And everyone knew the equation $E = mc^2$, even if they didn't quite grasp what it meant. Einstein became hugely popular with the American press corps and gained an avid—and growing—popular readership. In 1930, security staff at New York's American Museum of Natural History had to deal with a near riot when four thousand people tried to see a film offering to "demystify" Einstein's ideas.[2] In 1929, Sir Arthur Eddington—who was instrumental in confirming Einstein's theory of general relativity a decade earlier—gleefully wrote to Einstein, telling him that one of London's busiest shopping streets had been brought to a near standstill. Why? Because Selfridges, London's most prestigious department store, had displayed the text of one of Einstein's recent scientific papers in its windows, and Oxford Street was jampacked with people trying to read it.[3] Eddington himself went on to write what remains one of the most perceptive explanations of relativity,[4] offering a clear and reliable account of the scientific significance of these radical new ideas.

Einstein's influence continues to this day. In 2016, a team of scientists reported they had recorded two black holes colliding. This sound of a "fleeting chirp" from over a billion light-years away fulfilled the last prediction of Einstein's general theory of relativity.[5] Everything points to Einstein's scientific theories being here to stay and profoundly affecting the thinking of the next generation.

But beyond his scientific discoveries, what I have come to find really interesting is Einstein's *spiritual* significance. I write this book as someone who both encountered Einstein's ideas and discovered the intellectual and spiritual riches of the Christian faith at Oxford University. Although I will be aiming to give as reliable and accessible an account of Einstein's views on science as possible, I will also explore his ideas on religion and how he weaves these together. Yet perhaps more importantly, from my own personal perspective, I will also consider how his approach can be used by someone who, like me, wants to hold science and faith together, respecting their distinct identities yet finding a way of allowing them to enrich each other. My views are not the same as Einstein's, yet he has been an important influence in helping me navigate my way towards what I consider a workable and meaningful account of how this strange universe works and what it—and we—might mean. Einstein opens the way to trying to develop a theory of everything that matters.

I fell in love with science at about the age of ten.

My great-uncle, who was head of pathology at one of Ireland's leading teaching hospitals, gave me his old brass microscope when he retired. It turned out to be the gateway to a new world. I happily explored the small plants and cells I found in pond water through its lens. Then, having read some books about astronomy, I built myself a little telescope. On a cold, crisp winter's evening long ago, I turned it to look at the Milky Way and was overwhelmed by the number of stars I could now see. I was hooked and developed a love of nature that remains with me to this day. Einstein spoke of a sense of "rapturous amazement" at the beauty of nature.[6] I had not read Einstein at that stage, but I would have recognized what he was talking about immediately.

My first encounter with Einstein's scientific ideas dates from about 1966. In my enthusiasm to study science, I eagerly tried to absorb scientific works that I now realize were far too advanced for me. At the age of thirteen I plucked up the courage to ask one of my teachers to explain Einstein's theory of relativity to me. He loaned me one of his books to read. As I tried to take in Einstein's thought experiment about riding beams of light—to which we shall return later—I found myself struggling to grasp the points he was making. I realized that my mind needed to expand before I could make sense of Einstein. My problem as a thirteen-year-old was that I ended up reducing reality to what I could then cope with.

Happily, I was able to study Einstein in greater depth when I went to Oxford University in 1971 to study chemistry. The Oxford chemistry curriculum required students to specialize in one of a number of advanced subjects in our first year. I decided to focus on quantum theory, a field in which Einstein had made groundbreaking theoretical contributions while also asking some awkward questions. It was intellectually exhilarating. The lectures and seminars I attended opened my eyes to new ways of seeing our world. My research interests subsequently shifted to the biological sciences (my first Oxford doctorate was in molecular biophysics), yet I never lost interest in Einstein, whom I gradually came to see as a scientist whose interests extended beyond the natural sciences to embrace the fields of ethics, politics, and religion. As we shall see, Einstein is a role model for anyone trying to develop a "big picture" of reality that holds together multiple aspects of meaningful human existence.

Although I had no interest in religion as a younger person, seeing the natural sciences as the enemies of what I regarded as irrational superstition, I reconsidered this position during my first year at Oxford. I was aware that science had a wonderful capacity to explain the complexity of our universe—something Einstein explored in a series of groundbreaking scientific articles published during his *annus mirabilis* ("wonderful year") of 1905, to which we shall return later. Yet although I was thrilled at science's capacity to explain how things worked, it did not seem to be able to address deeper human longings and questions about meaning and purpose.

Many philosophers have explored this important point. Karl Popper, the great philosopher of science, spoke of "ultimate questions" dealing with value and meaning. These are important questions, affecting the lives of most human beings. Yet science cannot, by using its legitimate methods, provide answers to them. The Spanish philosopher José Ortega y Gasset puts his finger on the issue neatly: "Scientific truth is exact, but it is incomplete."[7] If we want to have a "big picture" of life, we are going to have to find some way of bringing together—and holding together—questions about how things *work* and what they *mean*.

Science has an important role to play in helping each of us construct our personal "big picture" or world-view. It can fill in part of that picture—but *only* part. As Einstein himself made clear, the sciences have their limits. They are not equipped to answer questions of value or meaning, and they are not *meant* to. If we are to make sense of our complex world, we need to use several ways of depicting it to help us appreciate its various aspects or components. Taken on their own, these aspects are like brushstrokes on a canvas. Yet when they are put together, they disclose a picture.

As a teenager, I assumed that my love for science required me to be an atheist. After all, science and religion were meant to be at war with each other—at least, according to the popular atheist tracts I had read. Yet it soon became clear to me that my teenage atheism was not adequately grounded in the evidence. It was a mere opinion on my part, which I had mistakenly assumed was a necessary outcome of reason and science. There were other options available. If I might borrow some words from the novelist Salman Rushdie, I discovered that "the idea of God" is both "a repository for our awestruck wonderment at life and an answer to the great questions of existence."[8]

Yet perhaps more importantly, I began to realize that Christianity, which I had dismissed as an outdated moral system with at best tenuous intellectual foundations, offered a way of seeing things—a "big picture"—that seemed to bring everything into a gratifyingly sharp focus. In developing my personal understanding of how science and faith could be held together in a productive and constructive manner, I found myself drawn to the approach of Charles A. Coulson, Oxford University's first professor of theoretical chemistry, who saw science and religious faith as offering complementary perspectives on our world.[9] Coulson set out what I found to be a deeply satisfying vision of reality that offered insights into the scientific process and its successes. At the same time he proposed a greater vision that allowed engagement with questions that were raised by science yet which lay beyond its capacity to answer.

I was interested to note that Coulson regularly cited Einstein in his exploration of the relation of science and faith. It was easy to see what Coulson had found in Einstein—a serious, reflective, and generous thinker, who sought to hold together what the philosopher John Dewey described as our "thoughts about the world" and our thoughts about "value and purpose."[10] Although my work at Oxford had focused on Einstein and quantum theory, it was not difficult to extend it to his other ideas. It is understandable that so many have focused on Einstein's scientific works. Yet Einstein was a remarkable human being, who tried to hold together his science, ethics, and religion in a coherent and meaningful way.

This book aims to explore these multiple aspects of Einstein's life and reflect on how he integrated them into a whole—a theory of everything that really matters. Einstein was an outstanding scientist who epitomized genius. Yet he was also a reflective human being who found himself caught up in the rise of Nazism in Germany and dragged into political and social debates that were not of his own making and not to his liking. The rise of Nazism seems to have caused Einstein to give careful thought to deeper issues of human meaning and values, which he believed might well be enriched by science but were nevertheless not disclosed or established by science.

As we shall see, one of Einstein's core ideas is that science is able to engage only part of our world. Physics is able to achieve a precise and accurate account of some aspects of our universe. Yet so much that is important cannot be expressed or formulated in this way. "How small a part of nature can thus be comprehended and expressed in an exact formulation, while all that is subtle and complex has to be excluded."[11] So much of what really matters to human beings seems to lie beyond the scope of the scientific method.

In this short work, I try to give a reliable account of both Einstein's scientific breakthroughs and his wider quest for a unified theory of everything. I do my best to explain his scientific breakthroughs as simply as possible, while referring readers to more advanced studies if they wish to consider these ideas further. Some readers may be surprised to find Einstein's religious views taken seriously, not least because they are so often dismissed and misunderstood. Yet they were an integral aspect of Einstein's identity, and he repeatedly emphasized the importance of holding science and religion together. Whether we agree with Einstein or not on these matters, he merits a respectful hearing on these points. Not only is he interesting; he also helps us work out how we can develop our own frameworks of meaning.

Einstein was a complex and nuanced thinker, making him vulnerable to ideologues who want to shoehorn his ideas into their own ways of thinking. Perhaps the most ridiculous of these distortions is the suggestion that Einstein's theory of relativity provides scientific justification for rejecting moral absolutes and adopting relativism.[12] Sadly, this remains an influential—yet quite mistaken—interpretation of Einstein. Many novelists of the 1920s mistakenly saw Einstein as confirming their moral relativism and weird ideas about time travel. Virginia Woolf, a leading member of London's smart Bloomsbury Group, concluded that, if Einstein was right, "we shall be able to foretell our own lives."[13]

Yet these popular misunderstandings of Einstein must not be allowed to prejudice his scientific achievement, nor do they invalidate informed attempts to open up deeper questions of meaning and value through engaging with him. Let's be clear about this from the outset: Einstein's theory of relativity does not endorse *relativism* but affirms a regular universe governed by laws. "My God created laws. . . . His universe is not ruled by wishful thinking but by immutable laws."[14] In a letter of 1921, noting some cultural misunderstandings of the scientific term *relativity*, Einstein suggested his approach was better described as a "Theory of Invariance" rather than a "Theory of Relativity."[15] We'll come back to this point later.

Einstein has also been conscripted by some propagandists as a mascot for their scientific atheism. Richard Dawkins's populist manifesto *The God Delusion* (2006) presents Einstein as a closet atheist who was "repeatedly indignant at the suggestion he was a theist."[16] Dawkins does not substantiate this incorrect assertion, offering instead a rather selective reading of some quotes from Einstein drawn from a secondary source.[17] What *really* annoyed Einstein, according to his own writings— which merit reading in their totality, rather than in selective snippets—was the repeated suggestion that he was an *atheist*, or being quoted by certain kinds of atheist writers as if he shared their views, particularly those he termed "fanatical atheists" with a "grudge against" traditional religion.[18]

It is easy to see how a hasty or superficial reading of Einstein—or an unwise dependence on secondary sources about Einstein—could lead someone to the view that he was an atheist. Einstein made it clear that he did not believe in a "personal God." Atheist apologists regularly interpret this to mean that he did not believe in God at all, overlooking Einstein's statements to the contrary. As Max Jammer, a personal friend of Einstein and professor of physics at Bar-Ilan University in Israel, points out in the most thorough and reliable examination of Einstein's religious views to date, Einstein "never considered his denial of a *personal God* as a denial of *God*" and was puzzled why anyone would even make this suggestion.[19] Einstein's ideas about God and religion don't fit our regular categories, and we need to listen to what he himself had to say about them, rather than forcing him into predetermined categories through selective quotation.

Yet before we begin to look at Einstein's careful calibration of the relation of science and religious faith, we need to consider his scientific theories and the difference that they have made to the way in which we—or at least, those in the know—think about our world. Since Einstein is so often said to have "overthrown" the views of Isaac Newton, it makes sense to begin our assessment of Einstein's significance by considering Newton's approach—often, though not entirely accurately, described as a "mechanical universe."

a revolution in science

chapter 2

THE OLD WORLD:
NEWTON'S CLOCKWORK
UNIVERSE

LET'S RETURN TO THAT HEADLINE in the *Times* of November 1919, announcing the verification of Einstein's theory of relativity: "Newtonian Ideas Overthrown." Just what were these ideas, and how were they "overthrown" by Einstein? What difference does this make? And how does their development help us understand how the natural sciences work? Let's begin by reflecting a little on what science is all about.

Thinking about Science

At one level, the natural sciences are about the patient and scrupulous accumulation of observations—whether we are talking about the movements of planets, the behavior of ants, or the patterns of rainfall in a region of the Andes. Yet science is about more than harvesting countless observations. It is a quest for understanding in which we try to discern the deeper patterns and structures of our universe that cause it to behave in certain ways. Science aims to go behind our observations and figure out what deeper truths underlie what we can see of our world.

It's an idea explored in the writings of the great Renaissance philosopher and scientist Francis Bacon (1561–1626), who compared "natural philosophers" to bees. (The word *scientist* did not come into use until the 1830s.) Why bees? Because bees do more than just gather pollen; they "transform and digest" what they collect and convert it into something new—honey. In the same way, the scientist both *gathers* observations and then *transforms* these by weaving them together into a *theory*—a way of understanding our world.

The word *theory* is widely used in science to refer to ways of understanding our world. It comes from the Greek word *theoria*, meaning "a way of beholding" or "contemplation." A theory is like a lens that allows us to see through and beyond the world of appearances to grasp something deeper and more fundamental that lies behind it. Like a set of spectacles, a theory allows us to see things in focus, helping us grasp how seemingly disconnected observations are actually part of something greater. Over time, certain theories gain widespread acceptance, becoming a settled part of the furniture of our minds. We get so used to these ideas that they seem natural and self-evidently true. So what happens if they turn out to be wrong? What if a familiar and trusted way of understanding ourselves and our world proves fallible and unreliable? What if a new theory comes along and is shown to be far superior to what we were used to?

Let's take an example—the movement of the planets. Even in the ancient world, people noticed that some of the objects in the night sky seemed to move around. Names were given to the five star-like objects that moved against the background of the fixed stars: Mercury, Venus, Mars, Jupiter, and Saturn. So what explained the movement of these "wandering stars"? In the second century, the astronomer and mathematician Ptolemy of Alexandria set out a way of understanding the heavenly bodies that would be accepted by most people for more than a thousand years. For Ptolemy, the sun, moon, and planets all revolved in circular orbits at different distances around the Earth.[1] It was a neat model, and it worked reasonably well, partly because, before the invention of the telescope in the sixteenth century, observations of planetary movements weren't very accurate.

But in the sixteenth century, awkward questions arose as increasingly accurate measurements of the movements of the planets became possible. The Polish astronomer Nicolaus Copernicus published a book arguing that the sun—not the Earth—stood at the center of the known universe. The Earth now had to be seen as simply one among many other planets. Copernicus's theory was wrong in several respects—for example, he believed that the planets revolved in *circular* orbits around the sun at *uniform* speeds. It needed correction before it could provide more reliable predictions of the movement of the planets against the fixed stars of the night sky.

Johannes Kepler corrected those errors through his close study of the movement of the planet Mars in the early seventeenth century. He pointed out that the Earth and other planets actually revolved in *elliptical* orbits around the sun at *variable* speeds. Yet Kepler couldn't explain why this was the case. In many ways, Kepler's achievement was simply to set out the rules that seemed to govern planetary motion. It remained a mystery why they should behave in this way in the first place. Kepler was unable to offer a theory that accounted for the existence and form of these rules. A bigger picture of the solar system was required to explain this.

As if on cue, the English mathematician and natural philosopher Isaac Newton (1643–1727) stepped up to the plate and ushered in a new and deeper way of understanding our universe.

Isaac Newton and the Laws of Nature

Newton set out his proposals for classical mechanics and gravitational theory in his *Philosophiae Naturalis Principia Mathematica* ("Mathematical Principles of Natural Philosophy"), now usually referred to simply as Newton's *Principia*, in 1687. His massive intellectual achievement was summed up by one of the leading poets of his age, Alexander Pope:

> *Nature and nature's laws lay hid in night;*
> *GOD said, Let Newton be! and all was light.*

So who was Newton? And what was his achievement? Newton developed most of his important work while he was professor of mathematics at Cambridge University. By careful observation, Newton set out a series of principles that governed the behavior of objects on Earth and then argued that these same principles applied to the motion of the moon around the Earth and the planets around the sun. This is the point made in the famous story of Newton and the apple.

This well-known story exists in several forms. The version of the story I was told back in the late 1950s was that Newton was sitting in his garden when an apple fell on his head—happily inflicting no permanent damage! Then, in a moment of devastating intellectual illumination, Newton invented his theory of gravity. It's a wonderful story. And like so many of these stories, there is an element of truth. Yet there's also a lot of exaggeration.

One account of this incident was written down by John Conduitt, who became Newton's assistant at the Royal Mint, which Newton directed in his later years:

> In the year 1666 [Newton] retired again from Cambridge to his mother in Lincolnshire. Whilst he was pensively meandering in a garden it came into his thought that the power of gravity (which brought an apple from a tree to the ground) was not limited to a certain distance from Earth, but that this power must extend much further than was usually thought. Why not as high as the Moon said he to himself & if so, that must influence her motion & perhaps retain her in her orbit, whereupon he fell a calculating what would be the effect of that supposition.[2]

This account was written down sixty years after the event. Scholars can hardly be blamed for wondering if Newton might have creatively embellished the story through retelling it over time, perhaps to conceal the fact that another British scientist—Robert Hooke—had developed a similar idea in the 1670s.[3]

The story of Newton's apple was given a surprising religious twist by the French antireligious philosopher Voltaire, who saw the apple as a symbol of science displacing religion as the cornerstone of human wisdom. Drawing on the tradition that the "forbidden fruit" in the Garden of Eden was an apple, Voltaire suggested that Newton's scientific discovery ushered in a new era in human history and self-understanding.[4]

Newton, of course, did not see his achievement in these terms. He rather saw himself as clarifying the deeper logic of the universe, discerning certain "laws of nature" that he considered to reflect the wisdom of God as creator. Newton argued that the mysterious and undetectable force he named *gravity* was the explanation for both an apple falling to the Earth and the moon orbiting around the Earth. Newton had no idea what caused gravity in the first place and refused to speculate about its origins. "It is enough that gravity does really exist, and . . . abundantly serves to account for all the motions of the celestial bodies."[5]

Initially, Newton's demonstration of the regularity of these laws of nature was seen as confirming the Christian belief in a God who had created an ordered universe and endowed humanity with the power of reason to discover those laws. Yet some were not so sure. Newton might well have shown that the universe was like a well-designed machine—cold, impersonal, and mechanical. But where was there any sense of beauty or joy? Furthermore, God now seemed to be pointless. Having constructed the universe and set it going, God is left without any significant role. God might retire or even die, but the universe would continue to function according to the laws by which God had caused it to function. Newton, perhaps unwittingly, had laid the groundwork for a self-sustaining and self-regulating universe, with no place for God.[6]

Newton on Space

Newton had demonstrated that the planets moved around the sun and the moon around the Earth according to certain "laws of nature" that could be expressed mathematically. But what did they move *through*? Newton introduced the idea of space—a vast, empty container that enclosed the sun, planets, and stars. This naturally raised some important questions, most obviously the question of what this "space" was made of. As with gravity, Newton refused to speculate on this question. Space was like an enormous box through which all objects moved in straight lines until some force caused their trajectory to curve. Like gravity, space could not itself be observed. Yet, again like gravity, it made sense of everything else that we observe. For Newton, both gravity and space were legitimate scientific inferences from an observable phenomenon to the unobservable entity that best explains it.[7]

Newton recognized that space and time were not things we observe directly but were rather inferences from those observations. However, many of his interpreters began to think of these as self-evident truths, things so obviously correct that they did not require justification or defense. Einstein would later express his concern about the ease with which such provisional concepts came to be seen as necessarily true and praised the German physicist Ernst Mach for insisting that scientific concepts arise out of experience:

> Concepts, which have proved useful in the ordering of things, easily acquire such a degree of authority over us that we forget their earthly origin. We take them as unchangeable givens. They come to bear the stamp of necessities of thought, of *a priori* givens. The path of

scientific advance is often made impassable for
a long time through such mistakes.[8]

So what was the significance of Newton's theories?
Basically, Newton was able to demonstrate that a vast
range of observational data could be explained on the
basis of a simple set of universal principles, such as
his laws of motion and the force of gravity. Newton's
theories worked well within the comfort zone of our
intuitions about the classical world. Yet when we move
away from this narrow world, we encounter the strange
world of the very small and very fast—the world which
Einstein later made his own.

Newton on Light

As we shall see, one of Einstein's most significant contributions to modern science concerned the nature of
light. So what did Newton have to say on this question?
In his *Opticks* (1704), Newton took the view that a beam
of light consisted of a series of rapidly moving small particles or "corpuscles" (from the Latin term *corpuscula*,
"small bodies"). Rays of light were actually beams of
tiny particles. Not everyone agreed with Newton here.
The great Dutch physicist Christiaan Huygens argued
that light was made up of waves and that some aspects
of its behavior could be explained using this model.

Newton, however, argued that the reflection of light by a mirror was just like throwing a ball at a wall and watching it bounce back.

Light Acting as Both Wave and Particle

Wave-like

particle-like

Newton's particle theory of light came to dominate physics during the eighteenth century. It led to some fascinating predictions, two of which are especially important for our purposes. First, if light consisted of a beam of particles, no matter how small, Newton's theories predicted that these particles would be influenced by gravity. They would travel in a straight line unless they were deflected by gravity. In 1804, the German mathematician Johann Georg von Soldner published a paper calculating the degree to which a beam of light would be deflected by the gravitational field of a star such as the sun.[9] Von Soldner noted that the predicted effect was far too small to be observed by the instruments of his time, so nobody followed through on this.

The second prediction was made by the English natural scientist John Michell in a paper presented to the Royal Society in 1783. Since light consisted of a beam of particles that would be attracted by the gravitational pull of a star, Michell argued that some stars might be so massive that their gravitational forces would prevent beams of light from ever leaving their surfaces. Like Newton's famous apple, these particles of light would simply fall to the ground. Michell thus proposed that "dark stars" existed, which could not be seen because light was unable to break free from the force of their gravity. Michell's calculations suggested that this would happen if the star was five hundred times the mass of our sun. Today, we know that Michell was right. We now call these dark stars "black holes."[10]

Yet during the nineteenth century, a growing body of experimental evidence suggested that light was better understood as a wave. In 1801, the English physicist Thomas Young devised a "double slit" experiment that suggested that light behaves like ripples or waves on a pond of water. By the 1870s, Newton's view that light consisted of particles had been abandoned in favor of a wave theory of light. It seemed to many that the issue had been resolved. But had it?

The Completeness of Nineteenth-Century Physics?

Many late nineteenth-century scientists believed that physics had finally sorted out all the great questions of the day and all that remained was to achieve greater precision in some measurements of natural properties. Writing in 1871, the great British physicist James Clerk Maxwell warned against any such complacency:

> This characteristic of modern experiments—
> that they consist principally of measurements—
> is so prominent, that the opinion seems to have
> got abroad, that in a few years all the great
> physical constants will have been approximately
> estimated, and that the only occupation which
> will then be left to men of science will be to
> carry on these measurements to another place
> of decimals.[11]

Maxwell had his doubts about this view. But there was no shortage of those who shared it. In 1894, the leading American physicist Albert A. Michelson gave an address marking the opening of a new physics laboratory at the University of Chicago. Michelson declared that the "future truths of Physical Science are to be looked for in the sixth place of decimals"—in other words, through more accurate measurements of what was already known.[12]

In 1888, the astronomer Simon Newcomb suggested that astronomy was now not so much about "the discovery of new things" as about "the elaboration of those already known." While a few new discoveries might be expected—such as observing hitherto unknown comets—Newcomb declared that we are "fast approaching the limits of our knowledge."[13] Robert Millikan, who went on to win the Nobel Prize in Physics in 1923, recalled his own days as a student in New York during the 1890s. He was constantly ridiculed by other students for his devotion to "a 'finished,' yes, a 'dead subject,' like physics." Why not work in a more interesting field that was going somewhere?[14]

Others, however, were much more cautious. It was true that the nineteenth century had made significant advances in the field of physics—above all in developing mechanical models to explain many aspects of the natural world. Yet enigmas and anomalies remained. Some observations just didn't fit neatly into the best theories of that age. The question was this: Were those enigmas and anomalies simply intellectual irritations that would soon be resolved? Or were they really signs that the existing scientific consensus was wrong and needed radical revision? Nobody really knew.

The historian Thomas Kuhn developed the idea of a "paradigm shift"—a radical change in the way in which scientists see our world.[15] He saw this kind of shift in the Copernican revolution of the sixteenth century, in which the view that the sun and planets revolved around the Earth was displaced by the very different view that the Earth was another planet revolving around the sun. Yet Kuhn's interest lay mainly in the way in which a dominant paradigm collapsed and gave way to something new. A tipping point was reached. It was realized that the old way of thinking just couldn't cope with the anomalies, opening the way to new ways of thinking that could make sense of these otherwise puzzling observations.

Let's look at one of these anomalies that troubled physicists at the end of the nineteenth century. Technically, this is known as the "advance of the perihelion of Mercury."

The Curious Behavior of the Planet Mercury

Newton's theory of planetary motion triumphantly explained the way the planets orbited the sun on the basis of his theory of universal gravitation and the laws of motion. It was widely seen as an intellectual marvel of the eighteenth century. The astronomer Johannes Kepler had early identified three basic laws of planetary motion. Yet these were simply summaries of the way the planets orbited around the sun. They did not explain *why* they behaved in this way. Newton's theory of gravity provided a robust and persuasive explanation of these observations, showing that planetary orbits reflected something deeper and more fundamental about our universe.

Newton's theory faced a major challenge with the discovery of the planet Uranus in March 1781. Although Uranus is actually visible to the naked eye, it had hitherto been assumed to be a star on account of the planet's faint image and its slow orbit around the sun. Once this "star" was recognized as a planet, its motion was studied carefully. It was found that the orbit of this new planet did not fit what was predicted on the basis of Newtonian mechanics. So what was wrong? Did the anomalous behavior of Uranus reveal a flaw in Newton's theory? Might Newtonianism have to be abandoned?

Yet astronomers quickly realized that there might be another explanation, which involved modifying Newton's theory, not abandoning it. What if there was an unknown planet beyond Uranus, whose gravitational pull distorted the orbit of Uranus? The calculations of the French astronomer Urbain Le Verrier and others suggested where this planet might be found. In 1846, the planet Neptune was observed. The anomalous behavior of Uranus was thus resolved without abandoning or modifying Newton's basic ideas. Many scientists of the nineteenth century saw the discovery of Neptune as *confirming* the reliability of Newtonianism.

Yet another planetary anomaly emerged during the later nineteenth century—this time concerning the small planet Mercury. Like all the planets, Mercury orbited the sun in an ellipse. The point at which it was closest to the sun—known as the perihelion—was discovered to move by a tiny, but observable, amount each year.[16] Why?

In 1859, Le Verrier, having successfully predicted the location of Neptune by calculating the impact of its gravitational field on the orbit of Uranus, proposed that this puzzling phenomenon could be explained by the gravitational pull of the other planets on Mercury.[17] His calculations suggested that most of this change could be explained in this way, but not all of it. One possible explanation of this discrepancy was the existence of a hitherto unknown planet of roughly half the mass of Mercury positioned closer to the sun. This would explain the observations on the basis of Newton's theory of gravity. This sparked a search for the proposed planet, which Le Verrier named Vulcan. It was never found.[18] At the end of the nineteenth century, the curious behavior of Mercury remained unexplained. What caused it?

On November 18, 1915, Albert Einstein reported to the Prussian Academy that the advancing perihelion of Mercury was explained precisely and persuasively by his new general theory of relativity. His new way of seeing the world had resolved this long-standing problem—along with many others. We therefore turn to consider the many new ideas developed by Einstein, beginning with the remarkable series of four articles published in 1905 that established his reputation as one of the most significant scientific thinkers of his age—despite the fact that he was working as a clerk in the patent office in the Swiss city of Bern, not as a research scientist at a leading Swiss university or scientific research institution. In what follows, we shall consider Einstein's early scientific career and focus on those four scientific papers that marked him out as stamped with that elusive quality of genius.

chapter 3

A SCIENTIFIC REVOLUTIONARY: EINSTEIN'S FOUR PAPERS OF 1905

Isaac Newton became a scientific celebrity in the *annus mirabilis*—the "wonderful year"—of 1665–1666. During that year, he developed the basic ideas of both differential and integral calculus and made the startling empirical discovery that white light is actually made up of colored rays. Indeed, Newton even provided a list of those "colors of the rainbow"—red, orange, yellow, green, blue, indigo, and violet. Yet this discovery was only the beginning of a remarkable career. Just over twenty years later, in 1687, Newton published the work widely known as his *Principia*, in which he set out his three laws of motion, the modern concepts of force and mass, and the new and deeply counterintuitive concept of universal gravitation.[1]

The acclamation of Newton as a genius was partly the result of a resurgent Britain wishing to impress its intellectual authority on the emerging world of science. Yet Newton's reputation did not rest entirely on nationalist grounds. His genius was held to lie in his capacity to penetrate beyond the world of experience to a deeper reality that lay behind it. The frontispiece to the 1729 English translation of Newton's *Principia* (originally written in Latin, which was just beginning to fall out of favor as the international language of scholarship) showed Newton sitting on the clouds of heaven, being personally instructed by the Muse of mathematics. Newton was a genius who could see beyond our world of appearances and discern a deeper structure, reflecting the mind of God. As the British astronomer royal Edmund Halley put it in a eulogy for Newton, his "sublime Intellect has allowed us to penetrate the dwellings of the Gods and to scale the heights of Heaven."[2]

Halley's comments help us grasp what was seen as special about Newton in particular and perhaps why there is such cultural reverence for the idea of a scientific genius in general. The genius is seen as standing above politics and other vested interests, possessing an almost religious authority on account of the quality of their vision of reality. Thomas Jefferson, the principal author of the Declaration of Independence and the third president of the United States, expressed the hope that a "natural aristocracy" based on "worth and genius" would emerge to replace the "artificial aristocracy" of wealth and power.[3]

It's not difficult to see why the concept of a genius was so attractive to many after the First World War, which ended in November 1918. There was an urgent need for rebuilding nations and putting things right after the catastrophe of a global war. Traditional sources of wisdom were seen to have failed. The world had emerged from a hitherto unimaginable trauma of destruction and devastation. Who could trust their political and social leaders after such an event? Perhaps a genius might show the world a better way. But who would this be?

Introducing Einstein

In 1905, Albert Einstein was working as a clerk in the Swiss patent office—or, to use its official name, the "Federal Office for Intellectual Property"—in the Swiss capital, Bern. He held no academic position in any prestigious research institute or university. His job was to assess patent applications. One of the most famous photographs of Einstein shows him standing at a lectern in room 86 on the third floor of the Federal Office for Intellectual Property, scrutinizing paperwork submitted by eager inventors hoping to secure a patent. The accounts of his time at the Swiss patent office suggest that he was diligent in his duties, and in 1906, after four years of service, he was promoted from technical expert (class III) to technical expert (class II).

His position at the Swiss patent office was intellectually undemanding, and Einstein found that he had time to work at the projects that really mattered to him—solving the riddles of physics that remained unsolved at the opening of the twentieth century.[4] As he later recalled, the patent office became his worldly cloister in which he "hatched [his] most beautiful ideas," often in discussion with fellow physicist Michele Besso, who worked in the same building. Although Einstein went on to gain worldwide fame after the end of the Great War for his general theory of relativity, and a Nobel Prize in 1921 for his explanation of the photoelectric effect, the roots of that later fame lay in a series of four groundbreaking articles he published in the leading scientific journal *Annalen der Physik* ("Annals of Physics") in 1905. We shall consider each of these articles in this chapter. But first, we need to tell the remarkable story of Einstein's rise to fame.[5]

Einstein was born to nonobservant Jewish parents at Ulm, in the kingdom of Württemberg, on March 14, 1879. Württemberg had recently become part of a unified Germany after the Franco-Prussian War in 1871. The summer after he was born, the family moved to Munich, the capital city of Bavaria, where Einstein began his education at the Luitpold Gymnasium. Later, following the failure of their electrical business, Einstein's parents moved to northern Italy while he continued his education in Munich.

Einstein's goal was to settle in Switzerland and train as a teacher in physics and mathematics at the Swiss Federal Polytechnic School (now known as the Swiss Federal Institute of Technology) in Zürich. His grades were not outstanding, and it was clear to Einstein that there was little likelihood he would go on to secure an academic position in Switzerland. Between May 1901 and January 1902, he taught at academies in Winterthur and Schaffhausen before moving to Bern, where he gave private lessons in mathematics and physics to make ends meet. The reasons for this move are not clear, although some suggest that it reflected Einstein's desire to avoid compulsory military service. Einstein had been stateless, having renounced his German citizenship in January 1896, but he acquired Swiss citizenship in 1901. In 1902, he found a relatively well-paid job as a technical assistant in the Swiss patent office. In January 1903, he married the Serbian mathematician Mileva Marić, who was a fellow student during his time at the

Zürich Polytechnic. The couple had two children: Hans Albert, born 1904, and Eduard, born 1910.

In the fall of 1900, while waiting for the outcome of his application for Swiss citizenship, Einstein used his spare time to keep up with his scientific reading. He was particularly influenced by the German physicist Ludwig Boltzmann, who explained the observed properties of gases as "discrete particles of definite size which move according to certain conditions." This "kinetic theory of gases" developed by Boltzmann emphasized that atoms were not some kind of hypothetical theoretical construction but were real objects. This ran counter to the dominant view of that time, which was forcefully expressed in the writings of Ernst Mach. Einstein later criticized Mach and his followers for allowing their scientific ideas to be determined by unacknowledged philosophical presuppositions, arguing that their prejudices against atomic theory were due to "their positivistic philosophical views."[6]

Boltzmann's ideas stimulated Einstein to write his first published scientific paper in 1901 on the implications of the well-known "capillary effect"—the ability of a liquid to flow in confined spaces without the assistance of gravity, or even in opposition to gravity. A familiar example of this is the phenomenon of "rising damp," in which water moves upwards in concrete or masonry, or the rising of sap in trees. But why did it happen in the first place?

Sadly, Einstein's paper—which was not particularly well written or argued—did not provide any convincing answers to this question.[7] Einstein's proposal for a correlation between the atomic weight of a liquid and the extent of its capillary action is no longer taken seriously. However, despite the shortcomings of his first scientific paper, Einstein now had the distinct advantage of having a published article to his name, which he circulated to academic institutions in the hope of finding a proper academic position.

In recent years, increased attention has been paid to this early article on account of a suspicion that Einstein might have had some unacknowledged assistance from Mileva Marić. At the time when Einstein was researching and writing this 1901 article, the couple were living together, although from mid-1900 until December 1902, they generally lived in different cities and sometimes even in different countries.

Einstein himself referred to the capillarity article of 1901 as "ours" in a letter to Marić, suggesting that she was somehow involved in its production. This has led some scholars to suggest that Marić was actually Einstein's collaborator and possibly the source of some of his ideas.[8] This point is important, given the remarkable series of papers that Einstein published in *Annalen der Physik* in 1905. Surely it would not be possible for one person to achieve such an astonishing feat, given the brilliance and originality of these articles?

This suggestion fits an influential media narrative, which (rightly) notes that male artists and academics of this period were prone to incorporate the ideas of female students or collaborators into their own work without due acknowledgment. A 2003 documentary titled *Einstein's Wife*, developed by Australian filmmaker Melsa Films, asserted that Marić was originally credited as a coauthor on several of these 1905 papers before her name was mysteriously removed from the final version of the texts. Was this a plot to disguise that she was actually a coauthor of these papers?

Following extensive scholarly criticism of this film,[9] this suggestion is no longer taken seriously, as it seems to rest on a series of misjudgments, including a mistranslation of some (questionable) comments by the Russian physicist Abram F. Joffe.[10] The evidence is now generally accepted to indicate that Marić did indeed support and encourage Einstein, helping him to track down information and perhaps occasionally checking his mathematics. She played a role similar to Michele Besso, whom Einstein valued as a "sounding board" for his ideas. Yet the core ideas were Einstein's own, even if he was wise enough to consult others in his quest to present them most effectively.[11]

So what were these four articles? And why did they have such an impact? In the sections that follow, we shall consider each one and explain its scientific importance.

March 1905—The Photoelectric Effect and the Nature of Light

What is now known as the "photoelectric effect" was first observed in 1887 by the German physicist Heinrich Hertz[12] and investigated more thoroughly later by Hertz's colleague Philipp Lenard in a series of studies, culminating in a major article of 1902.[13] Under the right circumstances, it was found that if a beam of light was shone on certain metals, the beam was able to eject electrons from the surface of those metals.

Perhaps this was not surprising. Classical physics recognized that every form of electromagnetic radiation transmitted energy, and it is quite easy to think of this energy pushing tiny particles of negative charge (i.e., electrons) off the surface of a metal, especially if these electrons were only loosely attached to that surface. It's a bit like throwing stones at a wall of rock. Sometimes the stones just bounce back, but sometimes they dislodge small pieces of rock as well. It all depends on how hard they are thrown and how many hit the wall at any one time.

As might be expected, Lenard's 1902 experiments found that the rate of emission of electrons from the surface of the metal was directly proportional to the intensity of light falling on it. The brighter the light, the more electrons were dislodged from the metal surface. Yet something else was observed that did not seem to fit existing theories.

Lenard also found that the brightness or intensity of the beam of light falling on a metal surface seemed to have no effect on the energy of these emitted electrons. The electrons emitted through exposure to a very bright light turned out to have the same energy as those emitted through exposure to a very dim light. It didn't really make sense. Common sense suggested that it was the brightness of the light that determined how much energy it transmitted. Furthermore, photoelectrons were emitted only if the frequency of the light exceeded a threshold frequency, which varied from one metal to another. (The frequency of light refers to the number of waves that pass a fixed point in a given period of time—typically one second.)

So why should the *color* of the light matter? Why did blue light seem more effective than red light? Surely it was the *brightness* of the light, not its color, that was really important? At that time, nobody thought that the color of light had anything to do with how much energy it transmitted. What really mattered, everyone thought, was its intensity. There was, of course, another explanation—that classical physics had missed something and Lenard's experiments were pointing towards a new way of thinking about the nature of light. So how could these observations be explained?

The Photoelectric Effect

An electron is emitted from metal when light hits it.

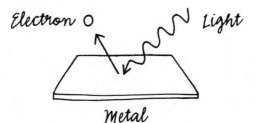

In his first article of 1905, Einstein proposed that according to the evidence, light seemed to be composed of particles (later named *photons*). He also proposed that each particle's energy was proportional to the frequency of the electromagnetic radiation of which it was part. In proposing this, Einstein drew on the theories of Max Planck set out in a publication of 1900 developing the hypothesis of the "quantization" of energy.[14] This hypothesis stated that the energy of an oscillator is not infinitely continuous but is made up of "packets" of fixed size. Planck introduced a fundamental constant, h (now known as "Planck's constant"), to refer to this basic unit of energy. For an oscillator of frequency v, the energy of the oscillator can be defined as hv.

According to Planck, energy is continuous but is actually made up of very small packets or discrete units. An analogy may be helpful in explaining this idea. It is like looking at a great sand dune in the African desert. From a distance, it seems smooth; on closer examination, you see it is made up of millions of small grains of sand. Or think of quantized energy being like the rungs of a ladder. If you want to climb the ladder, you can go up only in increments of one or more whole rungs. There are no intermediate levels.

Energy may *seem* to be continuous, but on closer examination, it is made up of tiny packets of energy. A quantum is the smallest bit of electromagnetic radiation that can be emitted—in some ways paralleling the idea of an atom as the smallest unit of matter (although this idea had to be abandoned in the 1930s, when Cockcroft and Walton demonstrated that they had split an atom). We may think that light is a continuous wave of electromagnetic radiation, but it is actually a stream of discrete packets of energy, now known as photons. Einstein considered the photo-electric effect to confirm Planck's ideas and to further undermine the increasingly beleaguered classical approach to the nature of light.

Einstein argued that the photoelectric effect was best understood in terms of a collision between an incoming particle-like bundle of energy and an electron close to the surface of the metal. The electron could be ejected from the metal only if the incoming photons possessed sufficient energy to dislodge this electron.

Einstein's theory allowed two of Lenard's otherwise puzzling observations to be explained:

1. The critical factor that determines whether an electron is ejected is not the intensity of the light but its frequency. Einstein here developed Planck's suggestion that the energy of an oscillator is directly proportional to its frequency.

2. The observed features of the photoelectric effect can be accounted for by assuming that the collision between the incoming photon and the metallic electron obeys the principle of the conservation of energy. If the energy of the incoming photon is less than a certain quantity (the "work function" of the metal in question), no electrons will be emitted, no matter how

intense the bombardment with photons. Above
this threshold, the kinetic energy of the emitted
electrons is directly proportional to the
frequency of the radiation.

Einstein's brilliant theoretical account for the photoelectric effect suggested that electromagnetic radiation
had to be considered as behaving as particles under
certain conditions. It met with intense opposition, not
least because it appeared to involve the abandonment
of the prevailing classical understanding of the total
exclusivity of waves and particles: something could be
one or the other but not both. It was not until 1915
that Einstein's approach began to achieve acceptance,
particularly on account of the meticulous research of
the American physicist Robert Millikan.[15]

This article led to Einstein's being awarded the Nobel
Prize in Physics in 1921 "for his services to Theoretical
Physics, and especially for his discovery of the law of the
photoelectric effect."

So why is Einstein's article on the photoelectric
effect so significant? What are its wider implications?
Historically, Einstein's article was of critical importance in developing the field of quantum mechanics.
Quantum mechanics arose in the aftermath of the
collapse of classical approaches to the nature of light

as a beam of particles (Newton) or as a wave motion, analogous to sound (James Clerk Maxwell). These two models of light were thought to be incompatible: if one was right, the other must be wrong.

Although Einstein was cautious in the way he stated it, his paper opened the way to thinking of light in terms of "packets of waves," an idea that could be interpreted as meaning that light could be seen as both a particle and a wave and that Einstein later referred to as the "wave-particle duality of light." The Danish physicist Niels Bohr, widely regarded as the founder of the "Copenhagen approach" to quantum theory, developed this idea using the idea of *complementarity*. For Bohr, the classical models of both waves and particles were required to explain the behavior of light and matter. Bohr's basic argument was that we need to use both particle and wave models of light to cope with the experimental observations, even though these are inconsistent and cannot easily be integrated into a larger picture of the nature of light.

Einstein, then, can be seen as one of the founders of quantum theory. Yet Einstein later did not like the way in which quantum theory developed during the 1920s, especially as it came to place an emphasis on probability. But by then, Einstein was famous for another reason—the theory of general relativity—to which we shall turn in the next chapter.

May 1905—Are Atoms Real?: Einstein on Brownian Motion

Words have histories, and the word *atom* is no exception. The word *atomos* (Greek: "something that cannot be divided") was introduced in the period of classical Greek philosophy by Democritus to refer to the fundamental components of reality. Atoms were seen as the basic building blocks of our world. Although there was sporadic discussion of the idea at many points in the history of science, it was not until the early nineteenth century that the idea began to solidify and become part of mainstream science.

This growing interest in "atomism"—the idea that the fabric of physical reality was made up of atoms—was largely due to the British chemist John Dalton. In his *New System of Chemical Philosophy* (1808), Dalton suggested that many of these chemical reactions could be understood in terms of atoms of different chemical elements combining with each other to produce new compounds. Dalton's atomic theory held that all chemical compounds were composed of combinations of atoms in defined ratios and that chemical reactions could be thought of as a rearrangement of the reacting atoms. Dalton was able to show that one tin atom will combine with either one or two oxygen atoms to form chemicals that we now know as tin oxide and tin dioxide, respectively. On the basis of his study of chemical reactions, Dalton also suggested that carbon dioxide was a compound made up of one carbon and two oxygen atoms. The Austrian physicist and philosopher Ludwig Boltzmann developed the ideas further, holding that gases—such as air—were made up of atoms and molecules.

This, however, proved controversial. Ernst Mach, Boltzmann's colleague at the University of Vienna, argued that atoms could not be seen or detected empirically; they were simply mental constructions that might be helpful in trying to make sense of our experience of the world. The idea of an atom was "a didactic utility"—a fiction that was nevertheless useful in deriving experimentally observable results. Although there were exceptions, the physics establishment around the year 1900 was generally of the view that atoms did not exist in reality.

All of this changed as a result of Einstein's second paper of 1905, which dealt with what is generally known as "Brownian motion"—the observation that very small particles of matter, when suspended in a liquid, do not remain stationary but move around randomly. The phenomenon was named after the Scottish botanist Robert Brown (1773–1858), who noticed that pollen behaved in this way when suspended in water and viewed through a microscope. The particles of pollen seemed to vibrate as if they were being shaken, and they moved within the liquid in irregular patterns.

Initially, Brown thought that the pollen moved because it was alive. After all, pollen was produced by flowers, which were living organisms. Brown eliminated this possibility by showing that the same pattern was observed when he used finely ground rock instead of pollen. "Extremely minute particles of solid matter, whether obtained from organic or inorganic substances, when suspended in pure water, or in some other aqueous fluids, exhibit motions for which I am unable to account."[16] Nobody else seemed able to make sense of his observations, which were easily reproduced in laboratories.[17]

Brownian Motion

The random motion that occurs when small particles are suspended in liquid.

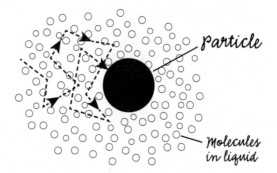

particle

molecules in liquid

This phenomenon of a constant erratic zigzag movement of small particles in a liquid was well known by 1905, when Einstein wrote his article. But what did it mean?[18] Einstein took the view that "the suspended particles perform an irregular movement . . . on account of the molecular movement of the liquid." In other words, the movement of the suspended particles in a liquid—like pollen in water—is due to the movement of the molecules of the liquid itself, which bump into the particles and thus cause them to behave erratically. Einstein was able to show how this random motion would be affected by, among other things, the temperature of the liquid and its viscosity.

Einstein explicitly drew on Boltzmann's "kinetic theory," which proposes that the temperature of a liquid is ultimately a measure of the rate of movement of its individual molecules, which could be compared to billiard balls moving around with increasing speed in a closed container. Raising the temperature of a liquid causes the molecules to move more quickly and hence to bump into any suspended particles with greater force.

On the basis of these assumptions, Einstein was able to derive an equation predicting the relationship between the extent of Brownian motion and the temperature of the liquid in which particles were suspended.

Broadly speaking, Einstein predicted that the motion of suspended particles increases with the temperature of the liquid, that it decreases with increasing viscosity of that liquid, and that it decreases with an increasing size of the suspended particles. Einstein knew he was taking a risk in proposing such a mathematically precise formulation, which could be verified or falsified by experiment. If his prediction of the amount of this movement was shown to be incorrect, a weighty argument would be provided against the molecular kinetic conception of heat—and the physical existence of atoms.

Several attempts were made in 1906 to confirm Einstein's approach, although the outcomes were initially confusing.[19] By late 1908, however, a steady stream of experimental results from French physicists emerged, which were strongly supportive of Einstein's theory.[20] For the French physicist Jean Perrin and others, the empirical verification of Einstein's mathematical formula implicitly affirmed the core assumption that lay behind it: liquids are made up of real physical molecules.

Einstein's greatest achievement in this paper was to show that Mach was wrong. It might not be possible to see atoms or molecules, but their real existence could be inferred from precisely the properties of particles suspended in liquid so carefully analyzed by Einstein in 1905. There were, of course, important parallels with

Isaac Newton's proposal of gravity. For Newton, gravity was a legitimate scientific inference from an observable phenomenon to the unobservable entity that best explains it. Similarly, atoms were not observed; their existence was inferred. Yet that process of inference was robust and generated further hypotheses that were open to empirical confirmation.

Following Einstein's article of 1905, the pace of exploration of the atomic hypothesis quickened. Interestingly, the notion of atoms was initially more readily accepted by chemists than by physicists. However, a number of factors catalyzed the further acceptance of the atomic hypothesis within physics. One of the most important was the discovery of the phenomenon of radioactive decay. This was discovered in 1896 by the French scientist Henri Becquerel, who noticed that chemicals containing uranium blackened photographic plates, as if they had been exposed to light. Further research in England by Ernest Rutherford and Frederick Soddy suggested that radioactive decay could result in the transmutation of one element to another. Radioactive elements, such as uranium and radium, were shown to emit what were initially termed alpha particles, beta particles, and gamma rays. The best explanation of these observations was that they involved change at the atomic level.[21]

We shall return to reflect on the importance of atoms when we consider another of Einstein's dramatic interventions of 1905, his formulation of perhaps the most iconic equation of the modern period: $E = mc^2$. But we must now turn to consider the next paper of 1905, in which he set out the basic themes of his theory of special relativity.

June 1905—Riding Beams of Light: Einstein on Special Relativity

Einstein's third article of 1905 set out his preliminary reflections on what we have come to know as relativity. Put simply, this is the basic idea that the fundamental laws and constants of physics are the same whether you are stationary or moving. Some will find this statement surprising in that they assume relativity is all about *relativism*—the idea that there are no absolutes and each of us can determine our own ideas. This is not what Einstein meant. In fact, the core assumption of Einstein's approach to relativity is that the laws of physics are universally true.

In an earlier chapter, we explored some aspects of Newton's classical mechanics, which set out the basic principles that govern the interaction of physical bodies. The crowning achievement of Newton's mechanics was a complete explanation of the workings of the solar system so that astronomers were able to predict with great accuracy and complete reliability such events as eclipses or the return of comets that lay decades or even centuries in the future. Yet Newton's mechanical philosophy dealt with large objects that were moving relatively slowly compared with the speed of light—such as planets orbiting the sun or apples falling to the earth. So what difference does it make if we consider the behavior of very small bodies—such as subatomic particles—moving at very high speeds?

In an autobiographical reflection written a month before he died in April 1955, Einstein recalled an early thought experiment that had intrigued him while he was a student at the cantonal high school in the Swiss city of Aarau: What would it be like to travel so quickly that it was possible to catch up with a beam of light?

> During this year in Aarau the following question came to me: if you were to chase a light wave with the speed of light, you would have a time independent wave field in front of you. Yet such a thing does not seem to exist! This was the first childish thought experiment that was related to the special theory of relativity.[22]

Einstein's point here is that if he were able to ride alongside a beam of light, it would appear as a stationary electromagnetic field. Yet surfing such a wave led to an impossible conclusion. How could an electromagnetic wave seem stationary—in effect, to become a frozen wave of light? Einstein knew that James Clerk Maxwell had shown that electromagnetic fields moved and vibrated. It had become clear that light was a good example of an electromagnetic wave. Yet Einstein's thought experiment suggested that under certain circumstances light stopped moving and vibrating. It didn't make sense. It was as if light had stopped being light. Something was wrong somewhere. But what? As Einstein remarked, the conflict between his thought experiment and Maxwell's equations caused him some degree of "psychic tension."

Einstein's third article of 1905 develops what was later termed the "theory of special relativity," which applies to objects that are moving in a straight line at constant speeds relative to each other. (Einstein's later theory of general relativity, proposed in 1915, applies to any form of motion. We shall consider this in the following chapter.)

Einstein's argument in this third paper of 1905 is based on two central assumptions: "the principle of relativity" and "the principle of the constancy of the velocity of light in a vacuum."

The "principle of the constancy of the velocity of light in a vacuum" holds that the speed of light in a vacuum has the same value, c, in all inertial frames of reference.

It's easy to accept—but, as Einstein would show, it leads to some counterintuitive consequences. The "principle of relativity" states that the laws of physics do not change, even for objects moving in inertial frames of reference. In other words, the same laws of physics govern the behavior both of stationary objects and objects that are moving with constant speeds in a straight line.

This point had been established by the great Italian astronomer and physicist Galileo Galilei in the seventeenth century—again, using a thought experiment.[23] In 1632, Galileo set out what we might call a classic version of relativity. Imagine you are standing on a dock, watching a ship sail past you at a constant speed. Now imagine that a sailor perched on the top of the ship's mast drops a rock. Where would this rock actually land? At the base of the mast? Or perhaps a short distance behind the mast, corresponding to the distance the ship had covered while the rock was falling? From the point of view of the sailor who dropped the rock and watches it as it falls, the rock falls straight downwards. And although common sense suggests that the rock ought to land *behind* the mast, it will actually land at the base of the mast.

Let's use a more modern analogy to make sure this point has been understood. Imagine you are on an airplane traveling at constant speed (typically about 500 miles per hour, or 223 meters per second). You drop the book you are reading onto the floor. Common sense suggests it ought to shoot past you towards the back of the plane. After all, the plane is moving forwards at 500 miles per hour. *But so is the book.*

As we all know from experience, the book falls directly to the floor. The book shares the same forward motion of the aircraft. Or, to use technical language, your book (and Galileo's sailor's rock) are both located within inertial frames. Both the book and the rock possess a forward motion as they are dropped on account of the inertial frames of the airplane or boat. This forward motion causes the book to land at your foot and the sailor's rock to land at the bottom of the mast—not where you might expect them to land on the basis of common sense.

Einstein develops aspects of his theory using thought experiments, which are rather more accessible than abstract mathematical formulation. In his later writings on relativity, Einstein asks us to imagine a real-life situation and think through its consequences. In what follows, I shall paraphrase Einstein at points while retaining as much as possible of the core elements of his approach.[24] So let's think ourselves into the analogy he uses in developing his theory of special relativity. Read each of the four elements of the analogy carefully, and make sure you are comfortable with the points being made:

1. Imagine a very long train traveling along the tracks at constant speed. The people inside the train will naturally adjust to their situation and use their carriage as a frame of reference.

2. Now imagine there is an embankment next to the track, which the train will pass on its journey. There are some people standing on the embankment, waiting to see the train pass. They are stationary; the train is moving at a constant speed of 60 miles per hour.

3. Now imagine that one of the people on the train starts to walk down the length of the train, in the direction of travel, at 4 miles per hour. What is her speed?

4. The key point here is that different observers will give a different answer to this question. For her fellow passengers on the *train*, the person walking along the train is moving at 4 miles per hour. Yet to an observer on the *embankment*, she will be seen as moving at 64 miles per hour. Her motion can be seen from different reference points. In fact, if she were to stop moving within the train, an observer on the *embankment* would still consider her to be moving at 60 miles per hour. But to someone inside

the *train*, she would not be moving at all. She would be stationary.

Einstein uses this and other analogies to make the point that motion is relative, not absolute. Each observer uses a frame of reference to assess both the direction and the speed of movement so that the same movement will appear to be different from different standpoints.

Now that we feel comfortable with this analogy, it can be developed further. Einstein asks us to replace the idea of a human being moving along the train with that of a beam of light traveling in the same direction.

Special Relativity Illustrated by a Train in Motion

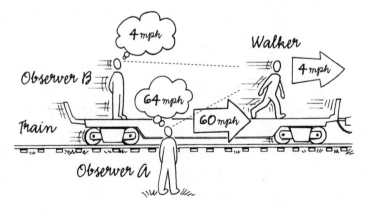

1. Suppose that someone on the train now turns on a flashlight and points its beam of light in the same direction as thexmoving train. The speed of light is c, and the speed of the train is v. To someone on the train, the light moves with speed c. But to someone on the *embankment*, watching what is happening inside the train from the outside, the total speed of the beam of light must be $c + v$, right? After all, they had already watched one of the passengers moving at 64 miles per hour, which is the combined speed of the passenger (4 miles per hour) and the train (60 miles per hour).

2. But that can't be right. The laws of physics do not change from one inertial frame to another, and the speed of light in a vacuum has the same value c in all inertial frames of reference. There's a contradiction here.

So how can this contradiction be resolved? Is there a problem with either or both of the two basic principles that Einstein presupposes? Or does the problem lie somewhere else? Let's allow Einstein to make the point himself:

If every ray of light is propagated relative to the embankment with the velocity c, then for this reason it would appear that another law of propagation of light must necessarily hold with respect to the carriage—a result contradictory to the principle of relativity. In view of this dilemma there appears to be nothing else for it than to abandon either the principle of relativity or the simple law of the propagation of light *in vacuo*.[25]

For Einstein, light travels at the same speed, no matter what the speed of its source of emission. So if the speed of light does not change as it moves through space and time, what other way is there of dealing with this dilemma? Einstein realized the need to rethink the relation of space and time. What the observer might see as changes in the speed of light actually reflect variations in what Einstein came to call "space-time." The solution had to lie in rethinking classical concepts of space and time. As a result, Einstein concluded that space and time must be seen as interwoven—a single continuum, known as space-time. This is not an easy point to grasp—which perhaps explains why nobody seems to have thought of it before Einstein. The same event can occur at different times for different observers. Time does not pass in the same way for everyone. Perhaps the best-known example of this is the "twin paradox," which concerns two identical twins, one of whom spends some time in a hypothetical spaceship traveling near the speed of light and returns to discover that his twin has aged much more than he has.[26]

Einstein's notion of relativity has, of course, been seriously misunderstood. As we noted earlier, the most significant and widespread misreading of Einstein presents him as offering scientific support for the idea of moral and cultural relativism. Some suggest that Einstein's theory of relativity is the outcome of his own radical political and social views—a symbol of "generational rebellion" against the fixed certainties of the past.[27] It is certainly true that many novelists and philosophers have argued that this is the case. For example, Gilbert Harman used Einstein's theory of special relativity to argue that "moral right and wrong . . . are always relative to a choice of moral framework."[28] Yet Einstein did not say that, and he certainly did not think that.

For Einstein, the physical laws underlying relativity are constant and universal. There is no sense in which they can be said to be "relative." Moral relativism has been around since the pre-Socratic philosophers in classical Greece, so its origins have nothing to do with Einstein. In any case, Einstein explicitly repudiated any connection between his theories and a system of morals or ethics. The great German physicist Arnold Sommerfeld declared that the essence of Einstein's theory was "not the relativity of space and time but rather the independence of the laws of nature from the viewpoint of the observer." Sommerfeld, however, felt that the phrase "theory of relativity" was an "unfortunate choice" on Einstein's part, given the way in which it was being misunderstood as somehow endorsing a "relativity of ethical conceptions."[29]

Richard Feynman, who won the Nobel Prize in Physics in 1965, was scathing about lazy and careless writers who were more interested in being trendy than in giving accurate accounts of Einstein's ideas. Feynman ridiculed "cocktail-party philosophers" who reduced Einstein's ideas to cultural slogans such as "all is relative" or "there are no absolutes."[30] As the (serious) philosopher Bertrand Russell remarked back in the 1920s, philosophers tended to interpret Einstein in the light of their own ideas and concluded that Einstein showed they were right all along.[31]

Yet Einstein's concept of relativity is the *result* of absolute laws, not their *denial*. Einstein believed he had restored order and coherence to our understanding of the universe, following the observation of anomalies that classic Newtonian mechanics couldn't explain. As he later pointed out, "Nearly every great advance in science arises from a crisis in the old theory, through an endeavour to find a way out of the difficulties created."[32] Instead of bending the rules to fit in these awkward observations, Einstein uncovered the deeper rules of our universe that ultimately explained these discrepancies. So let's get this straight: Einstein's theory of relativity has nothing to do with moral relativism. Einstein has simply been hijacked here by people who misread his ideas and then used them to justify their own moral and social views. Einstein's theory resolved an accumulation of scientific riddles that otherwise seemed insoluble. Perhaps his ideas seem to make little difference to our everyday lives—although global positioning devices have to use Einstein's theory to calculate our location. Einstein's achievement is to make it possible to develop a coherent understanding of our universe, something that lies at the heart of the scientific quest.

September 1905—$E = mc^2$: The Equivalence of Matter and Energy

Einstein's famous equation $E = mc^2$ states that the energy, E, of a physical system is numerically equal to the product of its mass, m, and the speed of light, c, squared. This assertion of "the equivalence of mass and energy" is widely seen as one of the most important principles underlying modern physics and was first set out by Einstein in the last of his articles of 1905, titled "Is the Inertia of a Body Dependent on Its Energy Content?" Although this short paper—a mere three pages long—does not explicitly use the equation $E = mc^2$, the basic idea is clearly formulated.

It was not, however, an original idea. In 1881, the British physicist J. J. Thomson proposed that a charged conductor in motion increases its mass in relation to its energy content. Perhaps more significantly, in 1904—the year before Einstein's paper was published—the Austrian physicist Fritz Hasenöhrl published an article in the same journal examining the properties of black-body radiation in a moving cavity and concluded that the radiation energy has an apparent mass associated with it.[33] While there were some problems with Hasenöhrl's calculations, the basic idea developed by Einstein the following year is unquestionably present in this earlier article.

But did Einstein prove that the equation $E = mc^2$ was right in 1905? Here, scholars are divided.[34] My own reading of the literature leads me to the conclusion that Einstein conceived the idea of "the equivalence of mass and energy" in the summer of 1905 but never managed to derive his idea from first principles. Like Eugene Hecht, I suspect that the way in which Einstein derives this idea in his 1905 article suggests "he already knew the answer he was looking for before he devised the rather elaborate scheme he used to reveal it."[35] Einstein returned to this question at least half a dozen times in the next twenty years, offering improved accounts of his basic idea and more rigorous justifications of his argument.[36]

My own view is that the idea of "the equivalence of mass and energy" seems to have been an intuition on Einstein's part, but an intuition rooted in a deep understanding of the physical world, so clearly displayed in his earlier articles of 1905. The philosopher of science Pierre Duhem wrote of experienced physicists possessing *le bon sens*, a rational faculty that allows us to intuit the truth of fundamental principles or axioms. Perhaps unkindly, Duhem suggested that German scientists lacked this capacity for intuition, which was better seen among their French counterparts![37]

Einstein hints at the influence of Duhem's approach on his research in a paper of 1918 celebrating the sixtieth birthday of Max Planck. Noting that the supreme task of the physicist is to search for general elementary laws that can be woven together to give a "picture of the world" (*Weltbild*), Einstein notes that "there is no logical path to these laws."[38] Rather, they arise through the "intuition, resting on a sympathetic understanding of experience." This seems to describe at least something of the strongly imaginative and intuitive process of reflection that lies behind Einstein's paper. He expressed much the same idea to Max Wertheimer towards the end of his life, remarking that he never thought primarily in the form of logical symbols or mathematical equations but rather in terms of images, feelings, and even music.[39]

In the end, what confirmed Einstein's formulation of the equivalence of mass and energy was not so much any arguments from first principles but experimental observations of the phenomenon. It is generally agreed that the experiment that confirmed Einstein's formula $E = mc^2$ was Cockcroft and Walton's famous "splitting of the atom" in 1932. This experiment, carried out at the Cavendish Laboratory in Cambridge, led to both Cockcroft and Walton being awarded the Nobel Prize in Physics in 1951 for their "pioneer work on the transmutation of atomic nuclei by artificially accelerated atomic particles."

So what did Cockcroft and Walton do? They constructed a piece of equipment that allowed them to accelerate hydrogen nuclei—protons—that had been created in a discharge tube down a long glass tube with a nearly 700,000-volt electrical gradient onto a plate of lithium metal. If a proton was absorbed by a lithium nucleus, it would become unstable and split into two atoms of helium, releasing energy in doing so. We can represent this as follows:

$$Li^3 + H \rightarrow He^2 + He^2$$

In a short letter to the editor of the journal *Nature*, Cockcroft and Walton explained how they had observed the production of alpha particles—helium nuclei— after bombarding lithium in this way.[40] They had split an atom.

Cockcroft and Walton did not mention Einstein's equation in their article, nor did they present their experiment as an attempted confirmation of Einstein's postulation of the equivalence of mass and energy. Nevertheless, this experiment is regularly cited as the first empirical confirmation of Einstein's equation. Why? The earlier equation I set out is not quite accurate, as it fails to mention that *energy was produced in this reaction*. When the energy of the two resulting alpha particles was measured, it was calculated that Einstein's theory predicted the conversion of the mass lost in this reaction to energy with an accuracy of 99.5%. A small amount of mass had been converted into a substantial amount of energy.[41]

So might there be some way to exploit the massive

potential of atomic nuclei to release energy in this way? In the year following Cockcroft and Walton's experiment, Einstein expressed skepticism about such a possibility, rightly noting that an enormous amount of energy needed to be used to accelerate protons to split atoms in this way, releasing a small amount of energy by comparison. Ernest Rutherford, widely considered to be Britain's leading physicist at this time, agreed, arguing that this "was a very poor and inefficient way of producing energy," even if the process was "scientifically interesting."[42] In part, the problem was that Cockcroft and Walton fired protons—positively charged hydrogen nuclei—into atomic nuclei, which were also positively charged. Inevitably, most of the protons were deflected through electrostatic repulsion. A lot of energy was wasted in accelerating protons that would not penetrate an atomic nucleus.

Nuclear Fission: The Splitting of an Atom

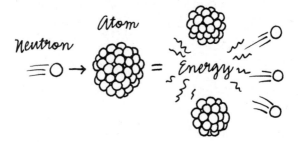

So what if neutrons—discovered by James Chadwick in 1932 and which have the same mass as protons but no electrical charge—were to be used instead? Might that make the process less wasteful and increase its potential to generate energy? In 1934, the Italian physicist Enrico Fermi announced that he had been able to create radioactive elements by bombarding atoms with slow neutrons in this way. Fermi's work with slow neutrons prepared the way for the discovery of nuclear fission, the key to extracting energy from nuclear reactions. Yet the key to the massive release of energy from atomic nuclei was realized to lie in a nuclear chain reaction, in which a neutron enters an atomic nucleus, causing it to split and release energy—and in doing so also release other neutrons, which could kick off the same process in other nuclei. But was this actually possible?

In early 1939—the year in which the Second World War broke out in Europe—German researchers demonstrated that uranium nuclei that were bombarded with neutrons released substantial amounts of energy in a process of fission, producing two new nuclei, called "fission fragments." It soon became clear that this process of fission caused neutrons to be emitted and that the neutrons from the fission of one uranium nucleus could thus cause the fission of other nuclei. In short, a nuclear chain reaction could be initiated, which would then sustain itself and release massive amounts of energy, which could be used in a controlled way to generate electrical power—or to create a bomb that could destroy entire cities.

Léo Szilárd, a Hungarian physicist who had emigrated to the United States in 1938, got wind of these developments and grasped their potential implications. Working alongside Enrico Fermi at Columbia University, Szilárd had realized that "secondary" neutrons were released as a result of the fission of uranium, setting in place the chain reaction that was essential for atomic reactors and atomic bombs. Worried that Nazi Germany might be able to establish an unassailable technological lead in these fields, with massive implications for the outcome of a future European war, Szilárd asked to meet Einstein and explain the significance of these developments.[43] They already knew each other from their Berlin days during the 1920s, when they had collaborated on the design of a new type of refrigerator, which failed commercially.

Einstein was alarmed. He agreed to sign a letter to President Franklin Delano Roosevelt, drafted by Szilárd, outlining the threat that the probable development of atomic weapons posed and recommending that the United States should invest in research in the field. The letter, sent to Roosevelt on August 2, 1939, did not recommend, as is sometimes suggested, that America develop its own atomic bomb. America was not then at war with Germany, and there was no expectation that this would happen. That situation changed, of course, in December 1941, following the Japanese attack on Pearl Harbor. Having declared war on Japan and its allies—including Nazi Germany—Roosevelt authorized the Manhattan Project, which led to the creation and use of the world's first working atomic bombs.

Einstein played no part in the Manhattan Project and seems to have been unaware of its existence or intended outcomes. The FBI regarded Einstein as unreliable politically and recommended that he should not have access to any sensitive information.[44] The use of atomic bombs thus came as a total and very unwelcome surprise to Einstein. He wrote a short piece titled "On My Participation in the Atom Bomb Project" for a Japanese magazine in 1952, making it clear that it was his belief that the Nazis might develop the atomic bomb first that motivated him to write to President Roosevelt, urging him to consider his options.[45]

So is Einstein to blame for the production of atomic weapons and the threat they are now seen to pose to the future of humanity? *Time* magazine certainly seemed to think so. Its front cover of July 1, 1946, shows Albert Einstein against a backdrop of a dark mushroom cloud, billowing above an atomic explosion. Emblazoned on the top of the mushroom cloud is the equation for which Einstein is best known: $E = mc^2$. Yet this is clearly unfair to Einstein, who had no idea that his postulation of the equivalence of mass and energy would lead others to produce and use viable weapons of mass destruction. Nobody in 1905 could have seen that Einstein's idea had military implications. Einstein can no more be blamed for nuclear weapons than Newton can be blamed for developing the equations that are now used to determine the trajectory of artillery shells or ballistic missiles.

Einstein's final article of 1905 raises difficult questions about the human use of scientific developments. We have developed nuclear and pathological weapons capable of wiping out entire populations, and we may quite possibly end up becoming the first species to bring about its own extinction. Einstein was acutely aware of the need for a broader context of discussion of scientific advances that took ethical issues seriously:

> By painful experience we have learnt that
> rational thinking does not suffice to solve
> the problems of our social life. Penetrating
> research and keen scientific work have
> often had tragic implications for mankind,

producing, on the one hand, inventions
which liberated man from exhausting
physical labor . . . but on the other hand . . .
creating the means for his own mass
destruction.[46]

In later chapters, we shall consider Einstein's ethical
and religious views, which inform his concerns about
how science was being used. In particular, we shall consider how he tried to hold together science, religion,
and ethics in a meaningful and workable manner. But
first, we need to ask what happened to Einstein after his
Wunderjahr of 1905.

chapter 4

THE THEORY OF GENERAL RELATIVITY: FINAL FORMULATION AND CONFIRMATION

EINSTEIN'S 1905 MAY INDEED have been a "wonder year," but his remarkable series of publications did not lead to offers of academic employment pouring in. In May 1905, Einstein submitted his doctoral thesis to the University of Zürich, titled "A New Determination of Molecular Dimensions." In July, he learned it had been accepted. Yet no academic appointment was forthcoming. Recognition and advancement of a more modest nature, however, lay to hand. In April 1906 Einstein was promoted to technical expert, second class, at the Bern patent office. It was not what he really wanted, but the increase in salary proved welcome.

Finally, the University of Zürich opened the door

to an academic career. In October 1909, Einstein was appointed as an adjunct professor in theoretical physics, allowing him to devote himself to the pursuit of science.[1] Two years later, he was appointed as full professor at the Karl-Ferdinand University in Prague, before returning to Switzerland in 1912 to take up a chair at the Eidgenössische Technische Hochschule (ETH) in Zürich. Meanwhile, his growing body of German admirers, headed by the physicist Max Planck and the physical chemist Walther Nernst, were working to bring Einstein into the heart of German-language culture and scientific research. Planck had been the associate editor of *Annalen der Physik* in 1905 and had been impressed by Einstein's articles of that year. (At that time, peer review was unusual, and Einstein sent his articles directly to Planck.) Those efforts proved successful. In 1914, Einstein was appointed director of the Kaiser Wilhelm Institute for Physics and professor in the University of Berlin.

Yet all was not well. Einstein's marriage came under stress shortly after the move to Berlin. Mileva and their two children returned to Zürich in 1914. In 1917, Einstein became ill and was cared for by his cousin, Elsa Loewenthal. In 1919, following the required five years of legal separation, Einstein divorced Mileva and married Elsa, who already had two daughters from her first marriage. The divorce agreement was unusual in that Einstein agreed to support Mileva and his two sons financially with immediate effect, as well as promising her the prize money, to be placed in trust for their two sons, if he were ever to be awarded a Nobel Prize.

In the meantime, European political tensions were rising. In July 1914, Einstein gave his first lecture to the Prussian Academy of Sciences. It was a momentous event in his personal history, marking the end of his transition from an academic outsider to a person of eminence. Yet that same month saw the Austro-Hungarian Empire issue an ultimatum to the Serbian government. The Austrians seemed willing to precipitate a general European war as a solution to their fragile international position. The outbreak of the First World War—then known as the "Great War"—in August 1914 marked the end of international intellectual and scientific collaboration.

Einstein was isolated in Germany during this period. He now had the time to try to resolve a major scientific problem that had preoccupied him since 1909: how to develop a generalization of his theory of special relativity. The scientific problem is not difficult to understand. The theory of special relativity that Einstein had set out in his landmark article of 1905 applied only to the relationship between a body at rest and a body moving at a constant velocity. The theory considered only the effects of relativity to an observer moving at constant speed. So what about bodies that were moving at *changing* velocities? And what about the influence of gravitational fields on space-time?

The Theory of General Relativity

Einstein was aware that Newton's law of universal gravitation appeared to be fundamentally incompatible with his own views on relativity. Newton had assumed that the force of gravity was generated purely by mass, whereas Einstein's equation $E = mc^2$ showed that all forms of energy had effective mass and must therefore also be sources of gravity. The theory of the equivalence of mass and energy raised important questions about how gravity was to be understood.

As we noted earlier, Newton had proposed gravity as a force between bodies as they moved through space, understanding space as a vast empty container. He had no idea what space was made of. Newton understood it in effect as a gigantic box through which objects moved in straight lines until some force—such as gravity—caused them to be deflected and move in curves. In the nineteenth century, the British physicists Michael Faraday and James Clerk Maxwell introduced the idea of electromagnetic fields. Maxwell was able to show that light "is an electromagnetic disturbance in the form of waves propagated through the electromagnetic field according to electromagnetic laws."[2] Einstein came to the view that, like electricity and magnetism, gravity was conveyed through a "gravitational field"—and, more radically, that this gravitational field is actually what Newton considered to be "space." Bodies do not move through space but through a gravitational field. Instead of thinking of space as a container through which planets move under the influence of gravity, we need to think of space itself as a gravitational field, which is distorted locally on account of the mass of stars.

On the basis of this approach, Einstein predicted the phenomenon of the gravitational dilation of time. The closer a body is to a large mass, with its substantial gravitational pull, the slower time runs for it. It is as if gravity is exercising an influence on the passage of time itself. This phenomenon is now well known and is important for the functioning of Global Positioning Systems (GPS), which rely on signals from satellites orbiting above the Earth to establish the observer's position. However, the atomic clocks in those satellites run 45 millionths of a second faster per day than clocks here on the surface of the Earth. Why? Because time passes at a different rate on the surface of the Earth due to the greater effect of the Earth's gravity.

Newton had understood matter to attract other matter across empty space. Einstein developed the quite different idea that matter distorts space-time. In effect, gravity causes space-time to bend around massive objects. Newton himself never applied his theory of universal gravitation to the behavior of light. However, as we noted earlier, scientists who held that light could be thought of as a beam of particles predicted that gravity would affect its passage through space. Two predictions are of especial interest. The first is John Michell's 1783 prediction of "dark stars" that could not be seen because light was unable to break free from the force of their gravity. The second is the 1804 prediction of Johann Georg von Soldner that a beam of light would be deflected by the gravitational field of a star, such as the sun. Soldner himself was able to calculate the extent of this deflection.

Einstein did not regard light as a beam of particles, which would be affected by gravity on account of their mass. His argument was somewhat different. Einstein's principle of the equivalence of mass and energy meant that light, on account of its enormous velocity, had an "effective mass." Newton thought light possessed mass; Einstein showed that it *behaved* as if it had mass and could therefore be attracted to other sources of mass—such as the sun and other stars. In effect, Einstein's theory of general relativity confirmed both these predictions but placed them on a different theoretical foundation.

The Trampoline
An illustration of how an object warps space and time.

So how can we visualize this changed understanding of gravity? The analogy I find most helpful is to imagine space-time as a trampoline. If you place a heavy object on a trampoline, it stretches the elastic fabric and makes it sag or bend. So imagine a trampoline on which someone has placed a lump of iron near its middle. Unsurprisingly, the fabric sags at that point. Now imagine that you roll a small ball across the trampoline fabric. It will move towards the lump of iron. Why? Because it is drawn to the iron? Or because it naturally follows the deformation in the shape of the fabric resulting from the weight of that iron? The second explanation is correct.

General relativity asks us to think of the sun and the planets warping space-time. The planets orbiting the sun are not really being pulled by the sun; they are actually following the curved space-time deformation caused by the sun. The astrophysicist John Archibald Wheeler summed up this view rather nicely when he remarked that "matter tells space-time how to curve, and space-time tells matter how to move."[3] Einstein thus converted gravitational physics into the geometry of space-time.

By November 1915, Einstein had worked out most of the details of how he could develop a more general theory of relativity. He had presented a draft and incomplete version of the general theory in the form of lectures the preceding summer at the University of Göttingen. In March 1916, Einstein submitted his paper, titled "The Foundation of the General Theory of Relativity," to the German scientific journal *Annalen der Physik*, which had published his landmark articles of 1905. The article was not as widely read as he might have hoped, due to the wartime conditions. Nevertheless, it was clear that the paper set out a comprehensive and sophisticated general theory, which was capable of being tested against observation.

So how are we to understand the relation of these

two theories of relativity—the special and the general? Einstein himself offered a neat explanation in an article titled "What Is the Theory of Relativity?" written in November 1919 in response to the massive public interest in his theories:

> The theory of relativity resembles a building consisting of two separate stories, the special theory and the general theory. The special theory, on which the general theory rests, applies to all physical phenomena with the exception of gravitation; the general theory provides the law of gravitation and its relations to the other forces of nature.[4]

One of the most significant aspects of Einstein's paper on "The Foundation of the General Theory of Relativity" was its specific predictions of what would be observed if the theory was correct. Einstein was absolutely clear that his theory had to be evaluated as a whole. As he remarked in 1919, its chief attraction lay in its logical completeness. "If a single one of the conclusions drawn from it proves wrong, it must be given up; to modify it without destroying the whole structure seems to be impossible." There were, in

Einstein's view, three such predictions that were open to testing:

1. A shifting of the perihelion of the planet Mercury, which arose on account of the planet's motion through space—that is, through a gravitational field—that was warped by the sun's enormous mass. Although this effect should be observed for all the planets, it would be most pronounced in the case of Mercury, which was so close to the sun's gravitational pull.
2. The phenomenon of gravitational lensing, in which the warping of space-time due to the gravitational influence of the sun causes light to bend.
3. The phenomenon of cosmological redshift. This prediction followed from Einstein's equivalence principle, noted by Einstein back in 1907.

Yet one prediction is strikingly absent from this short list—the expansion of the universe. Einstein's first cosmological solution of his field equations left him unsatisfied. Why? Because it indicated that the universe was expanding. In 1917, Einstein modified his equation, adding another term—the so-called "cosmological function"—in order to yield a static, unchanging universe. George Gamow later reported that Einstein thought this was the "biggest blunder" of his career (although the reliability of this recollection is questionable). In 1922 the Russian mathematician Alexander Friedmann used Einstein's equations to show that such a universe cannot be static. In 1929, the research of Edwin Powell Hubble suggested that the observational evidence was best explained by an expanding universe. Einstein's approach suggests that the universe expands not on account of the movement of galaxies but because space-time is expanding.

So why did Einstein not trust his original equations? Why introduce what his critics considered to be a "fudge factor" designed to fit the soon-to-be-abandoned model of a static and eternal universe? I have not found a convincing answer, although many suggestions have been made. Perhaps Einstein wanted his theory to fit in with physicists' thinking at the time; or maybe, as an admirer of the Dutch Jewish philosopher Spinoza, he genuinely believed that the universe was necessarily eternal. Yet with the benefit of hindsight, a century later, Einstein's "blunder" may actually have turned out to be an explanatory triumph. Einstein's cosmological function plays an important role in the dominant Lambda-CDM model of the origins of the universe and is seen to be critical in helping us understand the mysterious notion of "dark energy."[5]

If Einstein had predicted an expanding universe, his theory of relativity would have been sensationally confirmed by 1930. But let us return to the three predictions Einstein did make and consider their importance.

The first of these was more of a retrodiction than a prediction. In other words, it was something that was already *known* but which could not be *explained* by any existing theory. On November 18, 1915, Einstein presented a paper on the explanation of Mercury's advancing perihelion on the basis of his new general theory of relativity to the Prussian Academy of Science.[6] Although Einstein's formulation of the issue was later given added mathematical precision through the work of Karl Schwarzschild, his basic solution was perfectly adequate to show that this otherwise puzzling phenomenon could be convincingly explained in an elegant and persuasive way.

The third such prediction required significant technological advances if it was to be observed and so had to await future verification.[7] The absence of any evidence of a redshift in spectral lines was cited in British scientific circles in the next few years as a reason for having doubts about Einstein's theory.

The second prediction, however, could be confirmed

by existing instrumentation. All that was needed was a total solar eclipse so that the impact of the mass of the sun on starlight could be measured. How that happened is a story in itself.

The Confirmation of Einstein's Theory of General Relativity

Scientific teams had already attempted to verify Einstein's theories by observing star deflections during a solar eclipse. An American team from Lick Observatory in Mount Hamilton, California, had observed the total eclipse of June 8, 1918. Their conclusion was inconclusive, even pessimistic. However, the observatory had used low-quality equipment, and there were questions about the reliability of their results. There was clearly room for further investigation using more reliable equipment.

Einstein's scientific work was based in Germany and Switzerland and was published in the German language. All the available evidence suggests that British and American scientists knew little, if anything, of his ideas. Those who did were not particularly sympathetic to those ideas, which seemed to be in tension with those of leading English-speaking scientists, such as Lord Kelvin and Oliver Lodge. More seriously, the First World War had created major tensions within the international scientific community, with many British and American scientists expressing open hostility towards readmitting a defeated Germany into their circle.

Yet to everyone's surprise, Einstein found a champion in the British physicist Sir Arthur Eddington, who gained a reputation as one of the very few British scientists who was even interested in the theory of relativity, let alone understood it.[8] A story went around that Eddington was once complimented on being one of only three people in the world who understood the theory. Eddington didn't respond to this compliment. On being asked why not, he replied that he was trying to think of who the third person was.

Sir Frank Watson Dyson, the British astronomer royal, chaired the Joint Permanent Eclipse Committee (JPEC). On hearing of Einstein's prediction of the deflection of starlight in March 1917, Dyson decided that the eclipse on May 29, 1919, would be used to test this prediction. The star field near the eclipsed sun would be particularly rich, and the locations at which the total eclipse would last the longest were accessible to British astronomers. He recruited Eddington to lead the expedition.

As a Quaker, Eddington had been a conscientious objector during the First World War. This would normally have resulted in him being sent to a labor camp.

However, in a stroke of creative genius, Dyson seems to have managed to persuade the British authorities to see the eclipse expedition as an act of military service on Eddington's part. It was agreed that Sobral in Brazil and the island of Principe off the west coast of Africa were the best sites in terms of the totality of the eclipse, the anticipated weather, and the ease of access. Eddington himself led the Principe team. However, at the time of the eclipse, a significant degree of cloud cover caused some problems in taking good photographs. There were also some issues in the interpretation of the photographic plates.

The key question was the extent to which rays of starlight passing the sun would be deflected by the sun's gravitational field. In a paper written in advance of the expedition, Eddington suggests that there are three possible outcomes of the solar eclipse observations and considers their implications. Given their importance for Einstein's theory, we shall allow Eddington to set them out in his own words:

> With the enormous velocity of light only the sun would be able to produce an observable effect. The light ray from a star close to the sun should be slightly displaced; that might be tested next year when the totally eclipsed

sun would be in a field of bright stars. The experiment might lead to one of three results: (1) A deflection amounting to 1.75 seconds of arc at the limb of the sun would confirm Einstein, but would be twice as great as otherwise expected; (2) a deflection of 0.83 seconds would prove that light had weight, but would overthrow Einstein's theory; (3) the absence of a deflection would show that light, though possessing mass, had no weight, and hence the Newtonian law of proportionality between mass and weight would break down in another unexpected direction.[9]

Yet not all of Eddington's readers were persuaded that there were only three options. His critics suggested that there might be other possible explanations of any observed starlight deflections, including an uneven cooling of the Earth's upper atmosphere during the period of the eclipse. Why hadn't Eddington included these possibilities in his discussion?

In what many now see as a shrewd piece of media stage management, Dyson and Eddington arranged a dramatic event that they believed would decisively shape public opinion on Einstein's theories. On November 6, 1919, Dyson and Eddington were joined by the leading astronomer A. C. D. Crommelin at a much-hyped joint London meeting of the Royal Society and the Royal Astronomical Society. Dyson and Eddington declared to the gathered representatives of the British scientific community that expeditions to Africa and Brazil to observe the eclipse had indeed confirmed Einstein's prediction and ruled out its Newtonian counterpart.

This was presented as a stunning achievement that transformed our understanding of the universe. The *Times* headline, noted earlier, both captured and created a popular mood: "Revolution in Science. New Theory of the Universe. Newtonian Ideas Overthrown." The down-market *Daily Mail* made much the same point with its cheeky headline "Light Caught Bending." Other newspapers were caught up in this febrile mood. The *New York Times* declared that leading scientists were "More or Less Agog over Results of Eclipse Observations." A revolution had taken place in science; Einstein's theory was triumphant! However, the *New York Times* helpfully reassured its readers that there was no cause for immediate alarm; these discoveries did not "affect anything on this earth."[10] Einstein's brilliant theories had caught the imagination of the public through extensive media coverage, instantly propelling Einstein to international fame.

Yet scientific opinion remained divided—both

about the merits of Einstein's theory of relativity itself and about whether the 1919 eclipse observations really had confirmed it. The photographic plates resulting from these expeditions proved difficult to interpret at points. Some were so blurry that they had to be discarded.[11] This has led to suggestions—sometimes sensationalist—that Eddington and others may have discarded plates that did not fit in with their predetermined ideas, in effect fixing their results. However, a 1979 reexamination of the photographic plates shows that this suggestion does not appear to be substantiated.[12] Although many astronomers of the day, including Dyson, felt that these observations ought to be repeated at an eclipse predicted for 1922 before treating the case as closed, the 1919 plates were clear enough. Eddington's skills as a publicist had not interfered with his integrity as a scientist.

Many British scientists, however, were rather more disturbed by sensationalist newspaper reports of a "scientific revolution" that had "overthrown" Newton. This, it seemed to them, was simplistic and shallow journalism that failed to highlight the obvious and important *continuities* between Newton and Einstein. In April 1921, aware of this problem, Einstein himself gave an interview with the *New York Times* in which he emphasized that his theories were to be seen as an evolution that consolidated Newton's heritage rather than a revolution that discarded it:

> There has been a false opinion widely spread among the general public that the theory of relativity is to be taken as differing radically from the previous developments in physics. . . . The men who have laid the foundations of physics on which I build are Galileo, Newton, Maxwell, and Lorentz.[13]

Hoping to counter the loose and lazy language of a "revolution" in science, Einstein made it clear he considered his theories to be "the natural completion" of the work of earlier physicists, so that his new approach was to be seen as representing an "orderly transition" to a better way of seeing our world.[14] Einstein consistently emphasized his continuity with the theories of his precursors, such as Newton in the seventeenth century and James Clerk Maxwell in the nineteenth. Einstein thus did not regard his special theory of relativity as revolutionary but as a systematic development of earlier approaches. Yet the conventions and agendas of popular journalism led to his being portrayed as a revolutionary thinker who had overthrown an existing scientific orthodoxy in much the same way as Russian revolutionaries had toppled the czar and established a radically different world.

Yet while British and American readers lionized Einstein as a scientific genius, others took decidedly more negative attitudes—above all, back in Germany.

German Physics: The Campaign against Einstein

In 1919, Einstein was a German citizen occupying a senior position at the prestigious University of Berlin and was director of the Kaiser Wilhelm Institute for Physics. He was at the heart of the German scientific establishment and had become—largely through the influence of Eddington—an international celebrity. Such adulation for Einstein was echoed in the *Berliner Tageblatt*, which proclaimed that Einstein was a genius who had uncovered "a highest truth, beyond Galileo and Newton, beyond Kant." Finally, on December 14, 1919, the front page of the *Berliner Illustrirte Zeitung* was given over to an image of Einstein, proclaiming him to be "a new eminence in the history of the world." Einstein's research, it assured its readers, caused a "complete revolution in our understanding of Nature." Yet, as Einstein would discover, there were some unwelcome consequences of such uncritical adulation and publicity.

The surge in populist admiration for Einstein allowed his critics in the German scientific establishment to present him as an academic lightweight who indulged in both fraud and plagiarism to promote his flaky ideas.[15] Two events that took place in Germany in 1920 help us appreciate the hostility that was building up towards Einstein in the aftermath of his publications about relativity.[16] The campaign against Einstein was launched on August 6, 1920, with an inflammatory article by the populist journalist Paul Weyland in the *Tägliche Rundschau*, a Berlin daily newspaper, accusing Einstein of shameless self-promotion or stealing other scientists' ideas and presenting physics in purely theoretical terms, rather than as arising from experimental investigation.

To resist this contamination of German science, Weyland created the "Working Society of German Scientists for the Preservation of Pure Science" and organized a series of public lectures in the large auditorium of the Berlin Philharmonic Orchestra, capable of holding 1,600 people. In his opening lecture, Weyland ridiculed comparisons between Einstein and Copernicus, Kepler, or Newton. Surely Einstein should stop this shameless self-promotion in the media? After all, he seemed to know all the right people in the newspaper world.

Weyland here was hinting at something that would become increasingly significant in the 1920s: Einstein's

Jewish identity as something that was not properly German. (The newspapers that were most positive about Einstein's scientific achievements had close connections with Berlin's Jewish community.) The phrase "Jewish physics" began to be used, often contrasted unfavorably with "German physics." In 1937, Einstein's theory of relativity was criticized by Nazi sympathizers because of its alleged Jewish roots, which contaminated its scientific credentials.[17]

Tensions increased further as a result of a heated and inconclusive public debate, chaired by Max Planck, between Einstein and the distinguished physicist Philipp Lenard in September 1920. Although press reports emphasized the dispassionate and scholarly nature of this debate, it left Einstein deeply disturbed and wondering whether he had any place in the German academic world. Perhaps more worryingly, it suggested that his theory of relativity had become entangled with cultural and political agendas, above all with the question of whether there was a distinct "German physics," which was represented by Lenard—and not by Einstein. Was Einstein becoming un-German?

Although Einstein was clearly rattled by these developments, he did not consider leaving Germany during the 1920s.[18] Yet Einstein was not, and could not be, detached from the political and social controversies of his age and was actively affected by them. To understand scientists—including Einstein—we need to place them within the context of their social and political history. Einstein's theory of relativity was misunderstood by many in Germany during the 1920s. Some of these were genuine misunderstandings, while others seemed to be deliberate distortions aimed at discrediting Einstein in the eyes of certain social and political groups.[19]

The Nobel Prize in Physics

Einstein's brilliant theoretical work—evident in his articles of 1905—had clearly put him in the frame for a Nobel Prize. From 1910 to 1922, Einstein was nominated for the Nobel Prize in Physics every year except 1911 and 1915. Yet as events in Germany indicated, by 1920 Einstein had become controversial. His theory of relativity was too easily dismissed by critics as excessively theoretical. On the other hand, Einstein had become an international symbol of the explanatory and predictive power of science. Eventually, it was announced that the 1921 Nobel Prize in Physics would be awarded to Albert Einstein "for his services to Theoretical Physics, and especially for his discovery of the law of the photoelectric effect."

So why did it take so long for Einstein to be awarded

that prize? The Nobel Prize in Physics was awarded by the Royal Swedish Academy of Sciences, which based its decisions on the recommendations of the five members of the Nobel committee for physics and chemistry. In the early 1900s, this committee was dominated by a school of thought at Uppsala University that considered precise measurement the chief aim of their work. After all, theoretical developments ultimately rested on accurate measurements.

The 1907 Nobel Prize in Physics was given to the American physicist Albert Michelson for his "optical precision instruments and the spectroscopic and metrological investigations carried out with their aid." The Nobel Committee thus thought that Michelson's development of precise measuring instruments was more important than his theoretical developments based on those measurements!

In 1920, the Nobel Prize in Physics was awarded to the Swiss physicist Charles Édouard Guillaume for his discovery of Invar, a nickel-steel alloy that remained relatively unaffected by changes in its environment and thus allowed scientific instruments to make precision measurements with greater accuracy. Many saw it as an utterly bizarre choice. Guillaume was not an academic scientist at a prestigious university; he worked at the International Bureau of Weights and Measures. Yet understanding why Guillaume was awarded the 1920 prize helps us understand why Einstein was not.[20]

Unsurprisingly, Albert Einstein was nominated for the prize again in 1921. Allvar Gullstrand, professor of optics at Uppsala University, undertook a report on Einstein's contributions to relativity and gravitational theory and recommended rejection of any award.[21] The committee found itself deadlocked and ended up not awarding a Nobel Prize in Physics in 1921.

A chance conversation led Carl Wilhelm Oseen, a theoretical physicist from Uppsala, to the conclusion that Gullstrand simply did not understand Einstein's theoretical work or appreciate its significance. Yet there was nothing he could do about this; he was not a member of the committee that evaluated potential nominations. But in 1922, things changed. Oseen became a member of the nominating committee. Having himself grasped the importance of Einstein's work, he devised a strategy to get around Gullstrand's marked hostility towards Einstein, which was largely based on his theory of relativity—which had yet to be fully confirmed experimentally.

Why not instead award Einstein the Nobel Prize for his work on the photoelectric effect, which had been verified by experiment? In the end, this subtle compromise won the day. Einstein would receive a postdated

Nobel Prize for 1921 for his explanation of the photo-electric effect, "without taking into account the value that will be accorded your relativity and gravitation theories after these are confirmed in the future." Was the committee leaving the door open to a later award for the theory of relativity once this had been experimentally confirmed in the future?

The Nobel Prize increased still further Einstein's reputation and international appeal. Under the terms of Einstein's divorce agreement of 1919, his former wife, Mileva, was entitled to the financial—as opposed to reputational—benefits of the prize. As a result, she bought three apartment blocks in Zürich as an investment, living in one herself and renting out the other two. Einstein remained in Berlin, his reputation enhanced by this international recognition. However, his personal situation remained problematic, as growing anti-Semitism became an increasingly significant concern for him.

Einstein's Move to America

Following its defeat in the First World War, Germany faced a series of political and financial crises during the early 1920s as the postwar Weimar Republic found itself saddled with a massive reparations bill. It simply could not afford to pay it. The value of the German mark plummeted, causing massive inflation. In the first half of 1921, the US dollar was worth 90 German marks; by November 1923, the US dollar was worth 4,210,500,000,000 German marks. The value of savings went into freefall.

Anti-Semitism now began to emerge as a significant force in German politics. In an article of June 1921, Einstein spoke of his growing awareness of prejudice and discrimination against Jews in Germany. Yet far from causing Einstein to downplay his Jewish identity, it encouraged him to reaffirm and reassert it:

> Until seven years ago I lived in Switzerland, and as long as I lived there I was not aware of my Jewishness. . . . This changed as soon as I took up residence in Berlin. There I saw the plight of many young Jews. . . . These and similar experiences have awakened my Jewish national feelings.[22]

The Wall Street crash of October 1929 caused a worldwide slump of unprecedented severity, which proved fatal for the political stability of the Weimar Republic. The failure of a democratic government to come to grips with the crisis led to growing popular interest in authoritarian alternatives such as Communism and National Socialism (the latter being better known in its contracted form "Nazism"), which people hoped would provide firm leadership in the face of such crises. The National Socialist party leader was Adolf Hitler. Joseph Goebbels, under Hitler, translated the ideology of National Socialism into a popular movement, replete with a quasi-religious symbolism and strongly anti-Semitic mythology.

The emergence of a Nazi ideology seems to have spurred Einstein into developing his own creed—political, social, and religious. There is an interesting parallel here with the British novelist E. M. Forster, who penned *A Room with a View*. Writing in 1939, Forster remarked that the aggressive ideologies he saw in Nazi Germany and the Soviet Union forced him to develop his own beliefs in response. "There are so many militant creeds that, in self-defence, one has to formulate a creed of one's own."[23]

Einstein was rightly alarmed at the rise of Nazism. In 1930, he had taken the view that Hitler's growing influence reflected the economic crisis of the moment. Once economic conditions started to improve, he would no longer have the same appeal. Yet by late 1932, Einstein had realized that nothing could stop Hitler's rise to power. As a Jew, he knew that he and his family would no longer be safe in Germany. On December 12, 1932, Albert and Elsa Einstein left Berlin for the United States. Hitler was installed as the German chancellor a month later. German newspaper headlines made it clear Einstein would not be welcome if he returned: "Good News about Einstein—He Is Not Coming Back!"

Facing up to the inevitable, Einstein resigned from the Prussian Academy of Sciences and renounced his German citizenship. He eventually became an American citizen in 1940. His home would be the newly founded Institute for Advanced Study at Princeton University, where he remained for the remaining twenty-two years of his life. Yet Einstein's period in the United States, though unquestionably comfortable, did not lead to the scientific breakthroughs that some appear to have expected. His breakthroughs were made in the 1900s and 1910s. What followed was an attempt to consolidate and expand those earlier insights and a rigorous, yet ultimately fruitless, attempt to integrate the great theoretical insights of that time.

Einstein died at Princeton Hospital on April 18, 1955, of an abdominal aortic aneurysm. Perhaps aware of the risk of becoming the center of some kind of postmortem celebrity cult, Einstein stipulated that his body was to be cremated and his ashes scattered in the Delaware River. In at least one important respect, his last wishes were disregarded. In a bizarre twist of events, Einstein's brain was removed by Thomas Harvey, the pathologist on call that evening, and removed from the hospital without any authorization. Harvey also removed—again, without permission—Einstein's eyeballs and gave them to Henry Abrams, Einstein's ophthalmologist. Perhaps Harvey hoped to acquire fame and fortune through his illicit possession of Einstein's brain. In the end, however, the brain was returned to the pathology lab at Princeton in 1997.[24]

Einstein's Failed Quest for Theoretical Unity

Einstein's achievements were remarkable, from his decisive 1905 contribution to the formation of quantum mechanics to his development of the theory of general relativity in 1915. Yet Einstein saw his work as incomplete. Why? Because his quest for *die Einheitlichkeit*—the fundamental unity of all phenomena—remained unfulfilled. Einstein's writings of the 1910s show how he was driven by this quest. In a letter of November 1916 to the Dutch astronomer Willem de Sitter, Einstein remarked on his almost compulsive desire to find a unified view of reality: "I am driven by my yearning to generalize."[25]

His 1918 Berlin lecture to mark the sixtieth birthday of the great German physicist Max Planck is especially revealing here. For Einstein, the quest for a single "simplified and clear image of the world" was not merely the ultimate objective of science but also corresponded to a deep human psychological need to escape from the banality of everyday life into a purer world of objective perception and thought. The scientist's committed engagement to the study of the natural world thus required a "state of feeling" similar to that of a religious believer.

It is widely agreed that general relativity is at present the best generalized theory of gravitation and space-time structure. Yet although it can account for a remarkably wide range of phenomena, it is incomplete in that it ignores the quantum effects that govern the subatomic world. Most physicists adopt a pragmatic work-around to this problem, using general relativity to describe the large-scale phenomena of astronomy and cosmology and using quantum mechanics to account for the behavior of atoms and elementary particles. As these two realms are generally far removed from each other, this strategy works quite well in practice. But it is a theoretical fudge.

This approach is clearly unsatisfactory from a conceptual perspective. It is a work-around, not a solution. For a start, these two theoretical frameworks are quite different. General relativity has geometric precision and is deterministic; the world of quantum physics is shaped by uncertainties and is probabilistic—a feature that caused Einstein to have serious misgivings concerning its viability and that led to his famous (and often misunderstood) remark to the effect that God "does not play dice" with the universe.[26] More fundamentally, Einstein believed that there had to be a grander and more complete theory that enfolds both general relativity and quantum physics as special or limiting cases. Finding that grand theory that would embrace the "totality of empirical facts" was Einstein's life goal.

Einstein hoped to achieve a complete and coherent account of reality—a goal that some might see as philosophical rather than scientific.[27] Yet Einstein failed in this quest. Robert Oppenheimer, director of the Institute for Advanced Study at Princeton, where Einstein spent the final two decades of his career, suggested that Einstein's attempts to formulate such a unified theory were "a complete failure." Einstein was just "wasting his time."[28] Others, however, were more generous. The physicist Brian Greene believes that Einstein was simply ahead of his time. Mainstream physics of Einstein's time was simply not ready for the grand unified theory that Einstein believed lay tantalizingly within his grasp. "More than half a century later, his dream of a unified theory has become the Holy Grail of modern physics."[29] Einstein may have started from the wrong place, but he rightly grasped the possibility of holding together the complexities of the universe within a single grand theory.

Einstein's pursuit of a unified view of the world remains important beyond the world of physics. One of the central themes of this volume is the need to reflect on Einstein's belief that it was possible to hold together—if not to weave together into a coherent unity—his views on science, ethics, politics, and religion.[30] The search for a unified view of reality is not limited to physicists or cosmologists. We each, in our own way, try to weave together the threads of our beliefs and commitments in the hope of creating a coherent picture of reality—a *Weltbild*, to use the German term that Einstein favored.

Earlier Einstein scholars focused on the development of his scientific ideas and how they shaped and redirected scientific thinking. Yet more recent work has revealed the surprisingly large volume of material in which Einstein wrote about religious, ethical, and political issues, often reflecting on how these were to be related to each other. This growing interest in Einstein as a human being prompts two questions, both of which are explicitly addressed in this book. First, how did Einstein engage life's big questions, and what answers did he advocate? And second, irrespective of whether we agree with Einstein's views on science, religion, or ethics, can the way in which he tried to hold these together be of service to us as we try to develop our own view of reality, however modest and limited this might be?

We shall turn to consider these questions in the second part of this book.

a theory of everything

(that matters)

chapter 5

EINSTEIN AND
THE BIGGER PICTURE:
WEAVING THINGS TOGETHER

EINSTEIN WAS A "BIG PICTURE" THINKER, concerned about the ultimate nature of reality and our place within it.[1] In the late 1920s, Einstein began to set his sights on formulating a single grand theory that would weave together and unite relativity with quantum mechanics. He never succeeded, yet his quest for a "theory of everything" remains the holy grail of contemporary science. How can we weave various seemingly disconnected threads of thought together into a coherent whole?

In November 1944, Einstein wrote to Robert Thornton, who was hoping to launch a liberal arts program at the University of Puerto Rico, emphasizing the importance of the philosophy of science in generating a richer vision of our world. "So many people today," Einstein remarked, including professional scientists, seemed to be "like somebody who has seen thousands of trees but has never seen a forest."[2] For Einstein, it was important to develop a unified *Weltbild*—a coherent and comprehensive way of seeing our world—that allowed individual trees to be seen and appreciated for what they were while at the same time seeing them as part of something greater. It is a useful image, helping us appreciate the importance of *die Einheitlichkeit*—the fundamental unity of all phenomena—in Einstein's scientific and wider reflections.

Yet this unity of all things was more of an intuition than something that could be proved. The great German physicist Max Planck was aware of this aspect of Einstein's thought and considered it to be essential to the scientific method:

> As Einstein has said, you could not be a scientist if you did not know that the external world existed in reality—but that knowledge is not gained by any process of reasoning. It is a direct perception and therefore in its nature akin to what we call Faith. It is a metaphysical belief.[3]

The fundamental unity of phenomena thus turns out to be a philosophical or even theological belief, which provides both a motivation and justification for the scientific enterprise. It is not something that can be proved, but it nevertheless provides a working basis for the scientific project. For Planck, the scientist thus has faith in an unseen order of things and finds this both justified and reflected in the success of the sciences: "Anybody who has been seriously engaged in scientific work of any kind realizes that over the entrance to the gates of the temple of science are written the words: Ye must have faith. It is a quality which the scientists cannot dispense with."[4]

The ultimate test of a theory, according to Einstein, is the quality of its vision of reality—not simply in terms of the range of that vision, but in terms of the manner in which it is able to hold things together as a coherent and interconnected whole:

Creating a new theory is not like destroying an old barn and erecting a skyscraper in its place. It is rather like climbing a mountain, gaining new and wider views, discovering new connections between our starting point and its rich environment. But the point from which we started still exists and can be seen, although

it appears smaller and forms a tiny part of
our broad view gained by the mastery of the
obstacles on our way up.[5]

For Einstein, the best theory is able to accommo-
date the insights of older theories—such as Newton's
mechanics—but also offer a richer vision of our universe. The best theory weaves together what might
once have been seen as disconnected threads but that
can now be seen to be integral parts of the same "big
picture."[6] And, as Einstein rightly saw, this is an act
of *imagination* as much as of *understanding*. Einstein's
recognition of the importance of the imagination in sciences is well illustrated in his short essay "On Science":
"Imagination is more important than knowledge. For
knowledge is limited, whereas imagination embraces
the entire world, stimulating progress, giving birth to
evolution. It is, strictly speaking, a real factor in scientific research."[7]

Although such comments would seem strange to
more rationalist scientists, there is now a much wider
recognition within the scientific community of the
importance of the imagination in developing theories
and grasping the interconnectedness of reality.[8] Einstein
frequently remarked that logic had limited value in
creating links between theories and reality. Consider,

for example, the opening of "On Science": "All great achievements of science must start from intuitive knowledge. I believe in intuition and inspiration. . . . At times I feel certain I am right while not knowing the reason."[9] In his conversations with Max Wertheimer, Einstein suggested that he never thought in logical symbols or mathematical equations but rather found it more natural to use images, feelings, or even musical structures in his attempts to visualize the complex reality that could only be partly disclosed through science.[10] (Einstein once quipped that if he were not a physicist, he would probably be a musician.)

Einstein's views on the importance of such a "bigger picture" are dispersed throughout his writings and tend not to be presented in a systematic and comprehensive way. For Einstein, there is a single real world, one not subject to the control of the human mind but whose basic structures can be grasped, if only partially, by human reason through experimentation and reflection and expressed mathematically. This world can be investigated by various research methods and models, yielding results that sometimes appear inconsistent with each other but that, when rightly understood, can be seen as part of a unified account of reality. Einstein's genius was to find a way of holding together science, politics, and religion—in other words, to create a personal synthesis of everything that mattered. His own way of weaving them together is illustrative, not normative, showing us possibilities, not compelling us to follow him in every respect.

Einstein often wrote of the importance of developing a coherent *Weltbild*, a German term most naturally translated as "a picture of the world" or perhaps "a way of seeing the world." Einstein used the term *Weltbild* as a title for a collection of his essays touching on the many topics that he regarded as important and meaningful.[11] The English title of the book—"The World as I See It"—doesn't really express Einstein's basic idea that there is some deep, underlying unity or harmony to the universe. The point Einstein is making is not so much that human beings can "see" the world but that there is some fundamental unity and coherence to our complex world that allows it to be grasped, however imperfectly, by human reason and imagination. "The eternal mystery of the world is its comprehensibility."[12]

Einstein's writings reveal an intense commitment to ethical and political issues and a strong interest in religion as an appropriate response to the mystery of the world. Where some might try to compartmentalize these aspects of life or see them as mutually inconsistent, Einstein seems to have seen science, ethics, and religious faith as integral—yet different—aspects of an authentic human existence. It's not just Einstein's ideas on science, politics, religion, and ethics that matter; it's the fact that he was able to hold them together. Maybe he can help us do the same.

This quest for a unified view of the world lies behind what is generally agreed to be Einstein's greatest intellectual achievement—the theory of general relativity. Confronted with the incoherence of classic Newtonian mechanics and the electrodynamics of James Clerk Maxwell and Hendrik Lorentz, Einstein conceived a new way of understanding space and time that restored a significant degree of unity to physics.[13] Difficulties remained. Indeed, many would say that Einstein spent the last thirty years of his life trying and failing to find a way to combine gravity and electromagnetism into a single elegant theory. Yet Einstein was convinced that nature was essentially a unity and that a single theory could—and would—therefore eventually be found to represent it, despite his own personal failure to do this.

Einstein's quest for a "big picture," holding everything together coherently in a single whole, is neither exceptional nor unusual. In fact, it seems to be the default setting of the human mind. The pursuit of meaning seems "fundamental to human nature" and embraces "much of what we most prize and are willing to make sacrifices for," including such things as justice, art, and beauty.[14] Many literary figures, philosophers, and theologians would speak of some kind of instinct or intuition within human nature that both suggests there is such a unified vision of reality and stimulates us to find it. The philosopher David Hume, for whom Einstein had a particular respect, was always alert to the shortcomings of human reasoning and saw this quest emerging naturally from the deepest recesses of the human mind: "Reason cannot defend the principles which we need to steer us through our lives, and so nature takes over and engraves them on our minds. When rational justifications run out, we just go on in the way we find natural."[15]

Einstein's oft-repeated emphasis on the importance of a unitary view of our universe is, however, not universally accepted. Postmodern philosophers suggest that there is no "big picture," only a number of smaller pictures that are not necessarily connected with each other. Reality, it is suggested, is like a patchwork quilt, each panel of which is different and has little connection with its neighbors other than the threads inserted by the person making the quilt. This theme, deployed in works as diverse as the philosopher Nancy Cartwright's *The Dappled World* and the novelist Margaret Atwood's *Alias Grace*, suggests that there is no fundamental coherence to our universe. There are at best localized areas of pattern and meaning, none of which can claim exclusive authority.

The Parable of the Lion

Einstein held to a unified view of nature while emphasizing the limits placed upon humanity as we seek to grasp our vast universe in its totality. In 1914, Einstein wrote a letter to a friend using an analogy to help explain this limited grasp of reality: "Nature shows us only the tail of the lion. But I do not doubt that the lion belongs to it even though he cannot at once reveal himself because of his enormous size."[16] Einstein's parable of the lion remains illuminating in that it highlights the themes of both connectedness and distance in our attempts to grasp the deep structure of our universe and work out our place within it.

The first point Einstein's parable conveys is that what we observe of the universe is a manifestation of a far greater unseen reality that lies beyond our capacities to grasp and hold. In June 1927, Einstein expressed this in a different yet equally illuminating way at a fashionable Berlin dinner party. "Try and penetrate with our limited means the secrets of nature and you will find that, behind all the discernible concatenations, there remains something subtle, intangible, and inexplicable."[17] What we see is indeed connected to a greater reality, yet we can only know nature through observation and experimentation.

Einstein was a realist in a general sense of the term, holding that physics was essentially an attempt to construct a conceptual model of the real world. In 1924, he suggested that all natural scientists correctly saw themselves as trying to make sense of "something real that exists independently of their own thought and being."[18] Einstein often pointed out that the "real" is not given to us directly; what is given is our *experience*.[19] Science therefore cannot depict the "real" but offers a framework for accounting for our *experience* of the real. There is indeed a connection between our experiences and reality, but it is indirect.

This point was emphasized by many of Einstein's scientific colleagues around this time. Werner Heisenberg, noted for his contributions to the development of quantum theory, highlighted that we do not know nature *directly* but rather *indirectly* through our research tools. "What we observe is not nature itself, but nature *as it is disclosed by our methods of investigation*."[20] So if nature is a single totality, Heisenberg's observation leads us to the conclusion that different research methods lead to different *perspectives* on reality—not to a total view of reality itself. A way has to be found of integrating or weaving together these perspectives if we are to see reality comprehensively.

Einstein makes this point by emphasizing that we are dealing with the lion's tail, not the lion itself. Einstein spent the last thirty years of his life searching for the "tail" that would lead him to the "lion"—a "theory of everything" that would unite all the forces of the universe into a single equation. Might he be able to find a theory that accounts for "Newton's falling apple, the transmission of light and radio waves, the stars, and the composition of matter"? Yet such a unified theory always eluded him, even though he knew, in his heart of hearts, that it had to be there—somewhere.

Connecting Science and a Wider Reality

Some scientists see their laboratories as defining their worlds. Their concern lies entirely with what happens within their walls—with experiments, calculations, and theoretical reflection. They are aware there is a world beyond the laboratory walls, yet their intellectual and personal investment lies primarily in the pursuit of science. Everything else is of secondary importance.

Einstein was not that kind of scientist. His letters bear witness not merely to his wide interests and concerns—music, philosophy, politics, and religion are frequently mentioned—but to his desire to try to bring and hold the insights of these various fields together. He was a scientist who valued his connections with other worlds of thought, worlds in which he might not personally excel or possess an international reputation but that he nevertheless regarded as important and worthwhile. For example, Einstein loved music and frequently performed in public (although it is thought that no authentic recordings of any of these performances have survived). Yet his love for music was linked with a sense that certain composers were tuning in to something deeper about our world. He said, "Mozart's music is so pure and beautiful that I see it as a reflection of the inner beauty of the universe."[21] Indeed, he once suggested that Niels Bohr's model of the atom was "the highest form of musicality in the realm of physical thought."[22]

Might this help explain why some of Einstein's theoretical discoveries were triggered or stimulated by music? Consider, for example, Elsa Einstein's account of Einstein's discovery of the theory of general relativity, which she gave when dining with Charlie Chaplin in Beverly Hills in 1931.[23] Having told Elsa, sometime around 1915, that he had had a "marvelous idea," Einstein went to his piano and began to play, every now and then making a few notes. After half an hour of this, he went upstairs to his study to work on this idea in greater detail—and two weeks later completed his theory. This suggests that Einstein saw playing classical music as a brainstorming technique, capable of stimulating his thought and helping his emerging ideas coalesce.

Yet there is a deeper question here: whether there is some fundamental harmony between human thought and the deeper structures of the universe, an idea that was often discussed during the Renaissance in terms of the "music of the spheres." Although Einstein's scientific papers show his commitment to traditional scientific methods, he often spoke of theoretical physics as an attempt to uncover "the music of the spheres," which revealed a "pre-established harmony" within the fabric of the universe.[24] A very similar idea is set out in the great astronomer Johannes Kepler's *Harmonies of the World* (1619), which links music and astronomy in explaining the orbits of the planets.

Einstein suggested that, whereas Beethoven "created" his music, Mozart's "seemed to have been ever-present in the universe, waiting to be discovered by the master."[25] We see this thought in one of Einstein's scientific aphorisms: "In every naturalist there must be a kind of religious feeling; for he cannot imagine that the connections into which he sees have been thought of by him for the first time."[26] It's a thought we find in many early modern scientists such as Galileo and Kepler: the scientist is really uncovering the rationality of God, which is expressed in the universe. The scientist's thoughts about the universe somehow echo God's thoughts. As Stephen Hawking pointed out in an often-quoted aphorism, to find an answer to "why it is that we and the universe exist" is to "know the mind of God."[27]

So why did Einstein feel such an affinity with Mozart? Einstein's admiration for this great composer emerged during the 1880s, long before the great Mozart revival of the first years of the twentieth century. In his study of Einstein's core papers of 1905, the American physicist John S. Rigden suggested that Einstein's achievement in science was comparable to Mozart's in music.[28] Both were rooted in the assumptions and conventions of their day, yet both demonstrated an intuitive leap and a new way of seeing things that lay beyond the reach of those assumptions and conventions.

Einstein's brilliant theoretical resolution of the scientific dead ends and riddles of the opening decade of the twentieth century involved the discernment of something deeper and more fundamental that lay behind those puzzles and paradoxes. And when this deeper understanding was grasped, Rigden suggests, it allowed phenomena to be seen in a new way—and hence to be understood. Einstein did not see Mozart's music as a soothing backdrop against which he could think without distraction. In some way, it sensitized him to something more fundamental about our universe.

So how might we be sensitized so that we might also hear this harmony? Einstein is scathing toward those who are tone-deaf to the beauty of the universe and especially to the mathematical representations of its structures, which often possess an elegance that seems to be correlated with their truth. He singles out what he terms "fanatical atheists" for particular comment, remarking that "their grudge against traditional religion as the 'opium of the masses'" makes them unable to "hear the music of the spheres."[29]

Some may feel that such references to the "music of the spheres" suggest that Einstein has lapsed into some form of mysticism. There is no good reason to think that Einstein was a "mystic" in the general sense of the word.[30] However, he was certainly alert to the imaginative power of words. Like so many before him—including Newton and Kepler—he found the musical metaphor of "harmony" helpful in expressing and stimulating the study of the fundamental unity of the universe.[31] Perhaps that helps us understand the somewhat unusual way in which Einstein celebrated the confirmation of his theory of general relativity in November 1919: he bought himself a new violin.

A Theory of Everything That Matters

Einstein's writings of the 1930s and 1940s show an increasing interest in areas beyond the field of the natural sciences, including ethics, politics, and religion. In each case, he defines these in his own way and marks out his own distinct approaches. Although Einstein was open about his "inclination to a life of quiet contemplation,"[32] he knew there were issues that demanded his involvement and engagement. In one sense, there is little that is unusual in this. Most scientists have views on ethical, political, and religious issues, and many are active in debates and programs arising from these. What makes Einstein particularly interesting is his attempts to see these as *interconnected*. Again, this is by no means unique, but it is a characteristic feature of Einstein's *Weltbild*. All these perspectives are parts of a greater whole, woven together in the mind of the individual thinker but whose interconnection might not be apparent or demonstrable to an outside observer.

This is particularly evident in Einstein's 1949 essay "Why Socialism?"[33] Einstein here argued that the natural sciences cannot create moral goals, even though science may provide means by which those goals could be achieved. Such goals do not themselves arise as a result of scientific inquiry, yet science might help implement their application—for example, in the field of medicine. "Science . . . cannot create ends and, even less, instill them in human beings; science, at most, can supply the means by which to attain certain ends."[34] Einstein makes a moral case for socialism, fully aware that those moral norms cannot be established or confirmed by the natural sciences.

So are science and socialism incompatible or irreconcilable? After all, they are the outcomes of quite different processes of thinking, with quite different criteria of evaluation.[35] If someone is committed purely and totally to a scientific way of thinking, how can she be a socialist? Or, for that matter, have any political or ethical commitments, given Einstein's view that one cannot derive values through the scientific method? Einstein rightly saw that individuals and communities use different ways of thinking in order to build up a coherent view of the world, involving a complex interplay of the rational and the intuitive. Perhaps more importantly, he never drew the conclusion that, because science cannot establish moral norms and values, these are invalid or irrational *for that reason*.

Einstein developed much the same point ten years earlier in arguing for a constructive engagement between the natural sciences and religion.[36] Einstein here echoes

David Hume's point that a knowledge of how things function does not generate viable moral values:

> The scientific method can teach us nothing else beyond how facts are related to, and conditioned by, each other. The aspiration toward such objective knowledge belongs to the highest of which man is capable. . . . Yet it is equally clear that knowledge of what *is* does not open the door directly to what *should be*. One can have the clearest and most complete knowledge of what *is*, and yet not be able to deduct from that what should be the *goal* of our human aspirations.[37]

For Einstein, the fundamental beliefs that are "necessary and determinant for our conduct and judgments" cannot be developed or sustained in a "solid scientific way."

It is generally agreed that it is difficult to find a comprehensive ethical system either explicitly stated or implicitly assumed in Einstein's writings. Nevertheless, there is clearly an intuitive ethical vision that led Einstein to affirm the value of scientific research while criticizing some of its outcomes. Consider, for example, this powerful 1948 statement on the moral obligations of scientists:

Rational thinking does not suffice to solve the problems of our social life. Penetrating research and keen scientific work have often had tragic implications for mankind, producing, on the one hand, inventions which liberated man from exhausting physical labor . . . but on the other hand . . . creating the means for his own mass destruction. . . . We scientists, whose tragic destination has been to help in making the methods of annihilation more gruesome and more effective, must consider it our solemn and transcendent duty to do all in our power in preventing these weapons from being used for the brutal purpose for which they were invented.[38]

The fact that science enables us to do certain things does not make those things moral. A framework of values originating from outside science is needed to make such judgments. Scientists need to be able to make moral judgments, but science itself, according to Einstein, cannot provide us with moral guidance or establish our moral values.

Einstein thus argued that proper human functioning, both individual and communal, requires more than what a "purely rational conception of our existence" is able to offer. Yet discussing fundamental questions of meaning and value does not mean we have ceased to be rational. "Objective knowledge provides us with powerful instruments for the achievements of certain ends, but the ultimate goal itself and the longing to reach it must come from another source."[39]

Einstein's political commitments, which emerge from his ethical views, are similarly independent of science. His two core political values were the importance of intellectual solidarity among intellectuals as a means of reducing the threat of war, and the cultural Zionist movement as a model of a world community based on mutual solidarity.[40]*

So is there a single principle or central theme that holds these diverse ideas and values together? Or are they simply held together in Einstein's mind, shaped by his own personal history and narrative? To answer this question, we will need to look more closely at his quite distinct concept of religion and consider the role that this plays in his thought. Yet it will be helpful to introduce a philosophical voice at this stage that offers an analogy that helps us make sense of what Einstein is doing, while allowing us to develop our own approaches.

---

* It is important to note that Einstein's vision of Zionism did not include support for a Jewish state. Perhaps with Switzerland as a model, he envisaged the creation of a "national home" that would welcome Jewish immigrants and encourage Jewish cultural development while ensuring equal rights for Arabs.

Multiple Maps of Reality

How do we go about representing a complex reality? The English philosopher Mary Midgley developed the idea that we use multiple maps in our attempts to make as much sense as we can of our world and live meaningfully within it.[41] Consider an atlas, which provides us with many maps of a region—for example, North America or Europe. But why do we need so many maps? After all, there is only one planet Earth! Midgley's answer is simple: because different maps provide different information about the same reality.

Multiple Maps for the Same Reality
An illustration of how we understand complex realities.

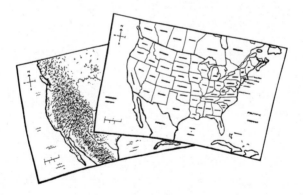

A physical map of Europe shows us the features of the landscape. A political map shows the borders of its nation states. Midgley's point is that no map shows everything. Each map is designed to answer a specific set of questions. What language is spoken here? Who rules this territory? Each map makes sense of the whole landscape only by answering certain questions about it and not others. To get an overall view of our complex reality, we refuse to rely on any single map. Instead, we try to find some way of bringing them all together so that their information can be harvested and used. It's not as if a physical map of North America makes a political map of that same region irrelevant. They answer different questions—and each of those questions is important.

Although Einstein does not use this specific imagery, it is helpful as an informing framework for his thought. Einstein is clear that the complexities of life demand answers to different questions. How does this work? What should I do? What does this mean? In effect, he is mapping different areas of life in terms of the questions that need to be asked and the resources we can use in answering those questions. And as we have seen, like any good scientist, Einstein knew that science had a limited capacity to answer questions beyond its own domain. We need scientific, ethical, political, social, and religious maps to help us navigate our way through the landscape of our world. No one map is able to answer all our questions.[42] We need to bring those answers together and try to weave them into a coherent unity.

We now will turn and consider Einstein's views on religion and see how they fit into this overall framework.

chapter 6

A "FIRM BELIEF IN A SUPERIOR MIND": EINSTEIN ON RELIGION

WHAT IS RELIGION? For some people, it's a boo word. It's all about outdated superstitions that have no place in the modern world. For others, it's what gives life meaning and purpose. Einstein uses the word *religion* in his own distinct way, which does not map easily onto what many people assume is its obvious meaning.[1] In this chapter, I will pay close and careful attention to Einstein's use of the term and try to tease out what he means by it before going on to ask how his ideas can inform or enrich our own ideas about religious faith.

From the outset, we need to be clear that Einstein was not religious *in the conventional sense of the word*. Though his Jewish identity came to be increasingly important to him during the 1930s, Einstein never attended any religious services. Perhaps most important of all, there was no religious ceremony to mark his passing. Einstein asked to be cremated immediately after his death and for his ashes to be scattered on the waters of the Delaware River.

Yet Einstein talked a lot about God. In a lecture on Einstein's intellectual achievements given at the Eidgenössischen Technischen Hochschule in Zürich in February 1979, Friedrich Dürrenmatt joked that "Einstein talked about God so much that I suspected he was a theologian in disguise."[2] In his published works, Einstein repeatedly and explicitly refers to an "intelligence," "mind," or "force" that lies behind or beyond the universe and identifies this with God. "This firm belief . . . in a superior mind that reveals itself in the world of experience, represents my conception of God."[3]

Einstein was explicit, however, that he did not believe in a "personal God." Although some read this to mean he did not believe in any kind of God,[4] Einstein was affirming his belief in some transcendent reality, which he was happy to designate "God," while making it clear that he did not understand this God as "personal." For Richard Dawkins, atheism is the belief that there is "nothing beyond the natural, physical world, no supernatural creative intelligence lurking behind the observable universe."[5] Einstein is clearly not an atheist on the basis of this definition.

While Einstein's God is impersonal, framed primarily in terms of the order and beauty of the universe, his concept of God cannot be reduced to a subjective feeling of awe. There is something beyond us that evokes a sense of awe, wonder, and mystery that Einstein sometimes describes as "cosmic religious feeling" and that he considered to be "the strongest and noblest motive for scientific research."[6]

At several points, Einstein specifically affirms his adherence to the concept of God associated with the seventeenth-century Dutch Jewish philosopher Baruch Spinoza—namely, an impersonal cosmic order. "I believe in Spinoza's God, who reveals himself in the orderly harmony of what exists, not in a God who concerns himself with the fates and actions of human beings."[7] Einstein studied Spinoza during his period in Bern and began to refer to him explicitly from 1920 on.[8]

So what did Einstein find attractive in Spinoza's notion of God? In a 1943 letter to Willy Aron, Einstein suggests that he was drawn to Spinoza because of their shared Jewish roots and culture.[9] Einstein is not, however, quite correct in stating that Spinoza held that God "reveals himself" in the order and structure of the universe. Spinoza did not think in terms of a self-revealing

God but rather tended to equate God with the "orderly harmony" of the world, which human beings can discover but which does not disclose itself.

As we have seen, Einstein was not an atheist. So was he a humanist? Alex Michalos, an academic at the University of Northern British Columbia, wondered how a "relatively hard-headed humanist" such as Einstein could take religion so seriously while conceding that it captured "some very important aspect of Einstein's own feelings and motives about his scientific work."[10] I assume that Michalos is here picking up on the fact that Einstein became an adviser to the First Humanist Society of New York in 1929. In the twenty-first century, the term *humanist* has indeed come to be understood to mean something like "atheist," "nonreligious," or "secular." However, this is not how the term was generally understood back in 1929, when, a year later, Charles Francis Potter, the founder of the First Humanist Society of New York, published the book *Humanism: A New Religion* in collaboration with his wife, Clara Cook Potter. This did not propose humanism as an *alternative to* religion but as a *form of* religion.

Such a "religious humanism" would have appealed to Einstein, as it echoed both his own positive views about religiosity and his more critical views of specific religious traditions, including Judaism and Christianity. Since then, more explicitly secular and atheist concepts of humanism have gained prominence in American and European public discourse so that the term *humanism* has come to mean a nonreligious, or even antireligious, worldview. This is not what the word meant at the time of the European Renaissance, and it was not what Einstein would have understood by the word in the 1920s and 1930s. Einstein often uses the term to refer to his concern for human freedom and well-being, particularly in the face of the rise of totalitarianism in Europe in the 1930s.

What Does Einstein Mean by *Religion*?

Einstein's many statements about religion have encouraged a selectivity on the part of his readers, especially when they have polemical agendas to pursue. It is not difficult, through highly selective citation of those statements, to present Einstein in three quite different ways: as a traditional religious thinker, as an atheist who had no place for religion, or as someone who was so confused on the matter that he is not worth taking seriously.[11] In what follows, the main elements of his approach will be considered without trying to shoehorn him into somebody else's categories.

The term *religion* is generally understood to mean

something like "believing in the supernatural." Yet scholars have long known that it is actually very difficult to define *religion* scientifically or culturally. For example, definitions of religion in terms of belief in gods or spiritual beings (such as Daniel Dennett's simple and inaccurate assertion that "a religion without *God* or *gods* is like a vertebrate without a backbone"[12]) have to come to terms with the problem of Buddhism, which just doesn't fit this description.[13] Many Eastern religions are actually philosophies of life, more akin to classical Stoicism than to what is normally understood by religion. In practice, many just assume that *religion* means what North Americans and Western Europeans think it means and are happy to leave it like that.

In clarifying what Einstein means by the term *religion*, I shall simply disregard what we think he *ought* to mean and concentrate on teasing out his own distinctive views. The philosopher Ludwig Wittgenstein famously argued that you should work out the meaning of a word by the way it is used. I will therefore offer a short synopsis of Einstein's use of the terms *religion* and *religious* and see where this leads us.[14] There are four main points that need to be made.

First, Einstein repeatedly refused to believe in a "personal God." This, as we noted earlier, is entirely consistent with his stated belief in Spinoza's conception of a God who is disclosed in the orderly harmony of what exists. Einstein here affirms belief in a specifically impersonal conception of God: "We followers of Spinoza see our God in the wonderful order and lawfulness of all that exists."[15] Einstein suggests that belief in a personal God is the "main source of the present-day conflicts between the spheres of religion and science." Why? Because the "doctrine of a personal God interfering in events" is not consistent with the "ordered regularity" of natural processes.[16] According to Einstein, God does not break the laws of nature.

Einstein expresses another concern about the idea of a personal God—namely, that it represents an "anthropomorphic conception of God," that is, a human construction of God that is simply a projection of aspects of human identity, which Einstein holds to be forbidden within Judaism on the basis of his reading of Exodus 20:4 ("You shall not make for yourself an image in the form of anything in heaven above or on the earth beneath or in the waters below").[17] But I wonder: Might Einstein have confused two distinct ideas here? After all, it is possible to hold a personal understanding of God without relying on anthropomorphic assumptions.

This concern helps us understand Einstein's hostility towards individual religions, which he tended to see as introducing ideas that were distortions of his own view of what religion ought to be like. He was particularly critical of the Christian Bible for its use of stories. How could the mystery of the universe be conveyed in this way? Einstein expressed this concern in a number of places, including the famous "God letter" of 1954, which speaks of the Bible as "a collection of venerable but still rather primitive legends." (This letter, written to the German philosopher Eric Gutkind shortly before Einstein's death, was auctioned for a record amount at Christie's in New York in December 2018.)[18] Yet such criticisms of Christianity and Judaism are part of a complex picture, and it is important to try to work out what Einstein's broader vision of religion is all about.

Second, Einstein sees religion as a response to something that ultimately lies beyond nature rather than a feeling of awe that arises in response to the vastness of the natural world. In his correspondence with Karl Eddi, Einstein remarked that, despite his own misgivings about the idea, a belief in a personal God was still "preferable to the lack of any transcendental outlook of life."[19] Religion is a recognition of this "transcendental" outlook, which Einstein often describes in terms of a force, mind, or reason that transcends both nature itself and the human capacity to grasp the natural world fully. "Try and penetrate with our limited means the secrets of nature and you will find that, behind all the discernible concatenations, there remains something subtle, intangible and inexplicable. Veneration for this force beyond anything that we can comprehend is my religion."[20]

Einstein is sometimes presented as offering a purely subjective experience of awe at the immensity of the universe, without affirming any transcendent foundation for such an experience.[21] There are indeed passages in which he speaks of the importance of such feelings of awe without tethering this to a transcendent referent, such as God. Yet these must be set alongside other passages in which Einstein specifically affirms a transcendent basis for such feelings or emotions—as, for example, in his statement of his "firm belief in a superior mind that reveals itself in the world of experience," noted earlier. Einstein's point is that this subjective response to the universe is not improperly invented by the observer but is properly grounded in something that lies beyond the observer.

Third, Einstein's view of God is not to be *identified* with that of Spinoza, particularly the latter's pantheism. Yes, there are continuities between Einstein and Spinoza, but there are also significant divergences. In an article in the *New York Times* in April 1929, Einstein declared that he was "fascinated by Spinoza's pantheism," while making it clear that he didn't think he could call himself a pantheist. Yet perhaps more notably, Spinoza stressed the importance of understanding our world and saw this rational comprehension of the universe as eliminating any sense of mystery in the face of nature. Such an attitude, he believed, would be eliminated once the intelligibility of the world had been grasped.[22]

Einstein, however, took a different view, constantly emphasizing the limits of our grasp of the rationality of the universe and the importance of a sense of mystery and wonder in the face of its vastness. Although he was not a mystic in any meaningful sense of the term,[23] Einstein was clear that a sense of "the mysterious" was the source of all true art and science, just as an "experience of mysteriousness" lay at the heart of religion. "What I see in Nature is a magnificent structure that we can comprehend only very imperfectly. . . . This is a genuinely religious feeling that has nothing to do with mysticism."[24] Where Spinoza saw the intelligibility of nature as confirmation of the competence of human reason, Einstein saw it as something that remained mysterious and puzzling. "I am imbued with the consciousness of the inability of the human mind to understand deeply the harmony of the Universe which we try to formulate as 'laws of nature.'"[25]

Fourth, Einstein's understanding of religion does not involve devotional practice or rituals. We have no record of Einstein ever participating in any Jewish worship. For many people, a religion involves or even requires participation in such observances. But not, it would seem, for Einstein. In explaining what he thought was so significant about Judaism, Einstein identified three key themes: "The pursuit of knowledge for its own sake, an almost fanatical love of justice, and the desire for personal independence—these are the features of the Jewish tradition which make me thank my stars that I belong to it."[26] There is no reference here to any aspect of Jewish religious practice. Furthermore, Einstein objected to forms of religious education that focused on religious ceremonies or rituals rather than on ethical values.

So how can we make sense of these four broad characteristics of Einstein's view of religion as well as the multiple scattered references to religion throughout his writings? My own view is that Einstein's general concept of religion, especially his notion of a "cosmic religious feeling" that is not tethered to any "anthropomorphic conception of God," is best understood as a *philosophy of religion*—that is to say, a set of ideas concerning a transcendent basis to the universe and the question of how we can know and represent it adequately without losing sight of its wonder and mystery.

As I read Einstein, I see a significant resonance with some eighteenth-century views on the "religion

of nature." "Religion is concerned with man's attitude toward nature at large, with the establishing of
ideals for the individual and communal life, and with
mutual human relationship."[27] This way of thinking
about religion is often believed to have arisen partly in
response to a dislike of organized religion, particularly
the tendency of religious institutions to get locked into
conflicts about their authority or privilege. This form
of religion was also skeptical about any formal belief
system; what really mattered was a sense of awe in the
presence of natural grandeur and beauty.

This "religion of nature" fits in well with Einstein's
repeated expressions of distaste for religious ceremonies
and specific religious doctrines. Historical religions,
such as Christianity and Judaism, included elements
that he found unnecessary or unacceptable—such as
the idea of a personal God who answered prayers, a set
of rituals and observances, or a foundational set of narratives or symbols. Einstein tended to regard historical
religions as incorporating mythical ideas that had no
place in his own understanding of genuine religion.[28]
This again echoes a leading theme of late seventeenth-
and early eighteenth-century European "religions of
nature."

Perhaps as a result of their suspicion of religious institutions, such "religions of nature" did not recognize any

privileged or authoritarian clergy or priests but rather saw every intelligent person who reflected on the beauty or order of the universe as a "priest." The best-known example of this is found in the writings of a great British contemporary of Newton, the chemist Robert Boyle. For Boyle, what he termed "natural philosophy" was a kind of spiritual exercise, intended to evoke respect or awe for nature.[29]

Let's return to Einstein's statement that "cosmic religious feeling is the strongest and noblest motive for scientific research."[30] This sets out clearly his core belief that religion (in his sense of the term) offers a fundamental motivation for scientific research that goes beyond a quest for understanding. This naturally leads us to consider how he understood the relationship of the natural sciences to religion.

Einstein on Science and Religion

Interest in exploring the relation of science and religion began to develop during the early modern period, as the notion of "natural philosophy" began to emerge as a field of research in its own right. During the seventeenth century, human curiosity—once seen as a vice—began to be seen instead as a virtue, capable of motivating people to explore the natural world and open up deeper questions of meaning. Natural philosophy both informed our understanding of the nature of God and was capable of inspiring moral and religious feelings.[31] By the time of Isaac Newton's death, there was widespread interest in considering how the natural sciences could inform Christian theology and vice versa.

Today, there are three main attitudes within Western culture concerning how science and religion relate to each other:[32]

1. *A War.* Science and religion are locked into conflict from which science alone will emerge victorious. There can be no meaningful conversations between science and religion, as this would amount to intellectual treason.

2. *A Silo.* Science and religion are totally different areas of human thought. They cannot—and anyway, *should* not—enter into discussion with each other. They are like insulated compartments of the human mind that do not interact with each other.

3. *A Dialogue.* Science and religion offer different perspectives on the great questions of life, which are capable of illuminating and informing each other. Such a conversation can be both constructive and critical at the same time.

I favor the third of these approaches. Richard Dawkins is an excellent example of the first such approach. Einstein doesn't really fit any of them, although most of his statements fit most easily within the second. Although he treats science and religion as quite different and noninteracting, he clearly believes that they could interact positively—for example, in exploring the relation of objective human knowledge and subjective human aspirations.[33]

Any attempt to discuss the relation of science and religion takes place amid the lengthening shadows of earlier controversies, which have since framed—and continue to frame—media discussions of the issue. During the 1930s and 1940s, Einstein was generally invited to discuss the relation of science and religion in the light of the prevailing assumption that these two aspects of human culture were at least in tension with each other, if not in a state of outright war. This perception was catalyzed by two factors. First, the publication of works such as Andrew Dickson White's *A History of the Warfare of Science with Theology in Christendom* (1896), which presented highly selective and unscholarly, inaccurate accounts of religious history to promote the idea that science and religion were permanently and necessarily at war with each other.[34] And second, the rise of religious fundamentalism in the United States during the 1920s, which often led to science being presented as the enemy of religion. The famous Scopes Trial of 1925 confirmed the growing popular impression that science and religion were incompatible. The hostility existed on both sides: there were some scientists who considered religion to be irrational and outdated, and some religious people who considered science to be intellectually and morally corrupting.

Although this media perception persists, modern scholarship has offered a more reliable account of the situation. In a series of important and influential historical studies, the Oxford scholar John Hedley Brooke argued that "serious scholarship in the history of science has revealed so extraordinarily rich and complex a relationship between science and religion in the past that general theses are difficult to sustain. The real lesson turns out to be the complexity."[35]

Brooke's analysis has found widespread support within the scholarly community. The Australian historian Peter Harrison has confirmed the outlines of Brooke's approach and pointed out that "study of the historical relations between science and religion does not reveal any simple pattern at all,"[36] such as the myth of the "conflict" narrative. It does, however, disclose a "general trend" that, for most of the time, religion has *facilitated* scientific inquiry. That's one of the reasons why I like the "dialogical" approach to science and faith. It allows for a serious discussion, with the potential for the enrichment of both science and religion.

Yet back in the 1930s, the popular perception was that science and religion were at war with each other. So how could Einstein suggest that there might exist some kind of positive and constructive relationship between them? In addressing the relation of science and faith in the United States during the 1930s and 1940s, Einstein had to engage some negative presuppositions about this relationship that were then seen as normative.

Einstein's approach was to treat science and religion as two distinct and different areas of human reflection, focusing on different aspects of our attitude to our universe. We want to know both how things work and what they mean. In a lecture given at Princeton in May 1939, Einstein considered the limits of rationalism in dealing with the big questions of life.[37] His core aim was to consider the relation of two different realms or modalities of human thought: science (facts) and religion (values). Einstein opens this lecture by setting the context, noting that during the nineteenth century it was "widely held that there was an irreconcilable conflict between knowledge and belief." This was linked with the rationalist perspective that "belief that did not itself rest on knowledge was superstition." In developing his own perspective, Einstein affirms that "convictions can best be supported with experience and clear thinking." This is completely consistent with Einstein's overall approach to science. But what happens if reason cannot establish the principles by which we should live our lives?

Einstein is clear that "those convictions which are necessary and determinant for our conduct and judgments cannot be found solely along this solid scientific way." What he has in mind here includes moral values and aspirations. When the scientific method is understood properly, these matters lie beyond its scope:

> The scientific method can teach us nothing beyond how facts are related to, and conditioned by, each other. . . . Knowledge of what *is* does not open the door directly to what *should be*. One can have the clearest and most complete knowledge of what *is*, and yet not be able to deduct from that what should be the *goal* of our human aspirations.[38]

Both the identity of that goal and the motivation to reach that goal "must come from another source."

Having briskly and efficiently shown the "limits of the purely rational conception of our existence,"

Einstein then moves on to explore how religion might transcend these limits placed on human reason. For Einstein, "the most important function which religion has to perform in the life of man" is that of disclosing our "fundamental ends and valuations" and to "set them fast in the emotional life of the individual." What he means by this is that religion discloses both the goals we should be pursuing in life and the ethical values that should guide us in this pursuit, as well as enabling these to become firmly anchored in an individual's personal existence.

In a second paper of 1941,[39] Einstein turned to deal more explicitly with the relation of science and religion. Science, he suggests, is fundamentally an attempt to "bring together by means of systematic thought the perceptible phenomena of this world into as thoroughgoing an association as possible."[40] It's a good summary of Einstein's attempt to go beyond the mere accumulation of observations to achieving an understanding of our world, perhaps extending to the quest for a unified view of reality. The goal of science is to discover rules that permit facts to be interconnected, while aiming to reduce "the connections discovered to the smallest possible number."[41] This "striving after the rational unification of the manifold" lay at the heart of Einstein's quest for the unification of reality.

Religion, however, is rather more difficult to define. Einstein suggests that it might be helpful to focus on the capacity of religion to liberate people from their selfish desires on the one hand and their conviction of their individual "overpowering meaningfulness" on the other. Why? Because this allows people to have a sense of their true goal and motivates them to achieve it.

If this is so, Einstein suggests, there can be no conflict between science and religion. Science cannot establish values; religion cannot deal with facts and their relationship. "Science can only ascertain what *is*, but not what *should be*."[42] Tensions can arise, Einstein suggests, when religion intervenes "into the sphere of science"—for example, in treating the Bible as a scientific text—or when science attempts to establish human "values and ends."

Thus far, Einstein appears to endorse a view similar to Stephen Jay Gould's idea of science and religion as distinct and divergent "nonoverlapping magisteria."[43] No meaningful or productive conversation is possible between them. Science and religion each occupy their own cultural silo or intellectual ghetto, from which outsiders are excluded. Einstein, however, declines to accept this conclusion. The realms of science and religion are indeed "clearly marked off from each other"; nevertheless, "there exist between the two strong reciprocal relationships and dependencies."[44] Science is ultimately motivated by religion; religion is informed by science. Einstein summarizes this in one of his best-known aphorisms: "Science without religion is lame, religion without science is blind."[45]

Einstein returned to the question of a possible conflict between science and religion in June 1948. After reiterating his belief that, when properly understood, science and religion could not be in conflict, he developed the idea that historical religions, such as Judaism and Christianity, tended to draw on certain "epics and myths" and that it was the mythical or symbolic content of religious traditions that was likely to come into conflict with science, particularly when religions made "dogmatically fixed statements on subjects which belong in the domain of science."[46]

Einstein does not elaborate on this point, but he appears to be criticizing those who treat the Bible as a scientific text—for example, in seeing it as providing information on the mechanism or chronology of the origins of the universe. This is consistent with the point made earlier: Einstein's concept of religion does not rest on institutions, defining narratives, or charismatic individuals but on a recognition of the wonder and mystery of our universe and the "superior mind" that lies behind it. He was critical of specific religions—including his own Judaism—for introducing ideas that he considered to be at best unnecessary and at worst inventions and delusions. In particular, as we noted earlier, he was critical of the idea of a *personal* God, seeing this as implying that such a God could and might interfere with or disregard the laws of nature.

The Problem of the "Now"

Einstein's analysis of the relationship of science and religion is, as we have noted, shaped by his own distinct understanding of religion, which is not easy to map onto the everyday realities that most people associate with this word. Yet the general points that underlie his discussion remain important. Science is not able to engage the human thirst for meaning and value in life; these need to be found elsewhere, and religion is able to engage these questions.

Earlier I suggested it was helpful to use the philosopher Mary Midgley's idea of multiple maps of a complex reality to reflect on the relationship of the natural sciences, ethics, politics, and religion. We are meaning-seeking beings. Science is one offspring of our perennial urge to make sense of our existence; religion is another. We might suggest that the natural sciences develop maps that deal with the question of how things function or the ways in which aspects of our universe are interconnected. Einstein argues that the sciences disclose "how facts are related to, and conditioned by, each other." That map is important. But we need more than just this map, as it does not disclose "those convictions which are necessary and determinant for our conduct and judgments." It is important to map the contours of the external world; we also need another map of the contours of our innermost longings, fears, and hopes. An objective account of our world does not easily or adequately engage our deepest existential questions.

To explore this further, it is helpful to reflect on Einstein's notion of space-time. As we have seen, Einstein's general theory of relativity sets out the notion of a four-dimensional space-time that warps and curves in response to mass or energy, thus causing the deflection of light beams as they pass by massive objects such as the sun. Although we intuitively think of space and time as quite separate notions, Einstein's theory invites us to see them as fused together to form a four-dimensional continuum. In effect, space and time are merged into the single idea of space-time. Its four-dimensional curvature can be described by Riemannian geometry. The amount of curvature in a region reflects the strength of the gravitational field at that point.

The Curvature of Space and Time
Space and time curve in response to the mass of an object.

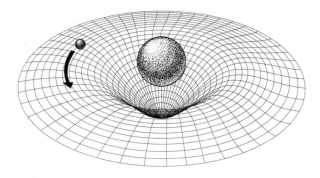

So where do *we* fit into this scheme of things? We can certainly think of ourselves as objects located in space-time. Yet one of the most distinctive features of human beings is that we try to make sense of our existence. Philosophers and theologians often make this point by emphasizing that human beings are both objects and subjects. Yes, we are located within the natural world, but we also think about this and try to work out how we fit into this world and what life is all about. The Austrian physicist Erwin Schrödinger raised concerns during the 1940s about what he described as the "removal of the subject from science." Physics has ignored scientists as subjective realities, even though their subjective experience is their essential link with the external world. The continuing interest in existentialism within Western culture reflects this anxiety that certain approaches to science dehumanize us by treating us as scientific objects rather than as thinking subjects.[47] A related point was made by Max Planck. Like Einstein, Planck held that there were limits to the capacity of science to fully understand our universe: "Science cannot solve the ultimate mystery of nature. And that is because, in the last analysis, we ourselves are part of nature and therefore part of the mystery that we are trying to solve."[48]

Subjectivity is one of the things that makes us special as human beings. We might indeed talk about "space" and "time," and some of us might even talk about "space-time"; yet we find it more natural to think in more subjective terms of "place" and "history." A place is a space that is special because of its associations or memories. "Home" is a good example of a place. It is about far more than a specific spatial location: it evokes a sense of belonging and identity. Home is where our heart is; it is an emotionally important idea, which transcends a mere location on a map.

Similarly, *history* designates far more than the mere passage of time. It's about things that have happened that matter and that affect us.[49] As *objects*, we exist in space-time; as *people*, we live in a place and in history. We want to reflect on our significance as individuals and so see the distinction between past, present, and future as more than changed coordinates in space-time. As human beings, we see this as a past in which we were not yet born, a present in which we live, and a future in which we are dead. That's why the "now"—this slice of space-time in which we exist as living and reflective beings—is so important to us. Yet, as Einstein realized, physics cannot make such a distinction, no matter how important this distinction might be to us.

Physics represents all the events experienced by a single person as a "world-line," a line in four-dimensional space-time. Yet there is nothing about any point on my world-line that singles it out as *my* "now." The German physicist Hermann Weyl thought this was a good thing, and it stopped physicists from becoming embroiled in pointless philosophical debates: "The objective world simply *is*; it does not *happen*. Only to the gaze of my consciousness, crawling upward along the life line of my body, does a section of this world come to life as a fleeting image in space which continuously changes in time."[50]

Other physicists were not so sure. Surely something was missing from this hyperobjective and totally impersonal account of human existence? Reading Weyl's views reminds me of a famous passage in C. S. Lewis's *Surprised by Joy*, which complains about the drabness of a rationalist world and its failure to connect with the deepest longings and questions of the human heart: "On the one side, a many-islanded sea of poetry and myth; on the other, a glib and shallow rationalism. Nearly all that I loved I believed to be imaginary; nearly all that I believed to be real I thought grim and meaningless."[51]

This problem was noted and addressed by Sir Arthur Eddington, who played such a critical role in confirming Einstein's theory of general relativity. In his 1928 classic *The Nature of the Physical World*, which introduced the phrase "Time's Arrow," Eddington noted how there was a mismatch between the pure objectivity of physics and the subjective experiential world of individuals: "Something must be added to the geometrical conceptions comprised in Minkowski's world before it becomes a complete picture of the world as we know it."[52] For Eddington, this "picture as it stands is entirely adequate to represent those primary laws of Nature"; it is not, however, adequate to engage our inner perceptions of the passage of time or other subjective concerns.

Einstein discussed this problem with the philosopher Rudolf Carnap, who worked at the Institute for Advanced Studies at Princeton for two years from September 1952, thus overlapping with two of the final three years of Einstein's life. As Carnap later recalled in his personal memoirs, Einstein was concerned about the problem of "now." Human beings regarded the present as distinct from the past and the future and considered it to be special, but this distinction could not be sustained on the basics of physics alone. "Einstein thought that these scientific descriptions cannot possibly satisfy our human needs; that there is something essential about the 'Now' that is simply beyond the reach of science."[53]

Einstein's understanding of the relationship of science and religion can be seen as an attempt to integrate (or at least to hold together) the objective and subjective aspects of human existence, recognizing that both are important parts of a larger account of life. Some atheists may regard this as a ridiculous suggestion; others may welcome it while making the point that its generality makes it difficult to develop. Not many share Einstein's idiosyncratic concept of religion. Surely there is a need to place Einstein in dialogue with specific communities of faith and ask where those conversations might go? In doing this, nobody is suggesting that Einstein is a member of such communities, simply that a dialogue might have some worthwhile outcomes.

We shall develop this point in the concluding chapter of this work, which asks a question that both needs to be asked and can be answered: What might a Christian learn from reading Einstein? How does Einstein inform and engage a Christian "big picture" of our world? In asking this question, we move beyond offering an analysis of Einstein's ideas and achievements and explore how Christians might weave his ideas into their own understandings of our world.

chapter 7

GOD AND
A SCIENTIFIC UNIVERSE:
TOWARDS A CHRISTIAN
READING OF EINSTEIN

BY ANY STANDARDS, Einstein was a brilliant thinker. Yet he wasn't right about everything. His views on quantum mechanics—a field he helped to pioneer—are now generally regarded as wrong. Perhaps more problematically, his social views do not sit easily with today's moral values. In 1922, Einstein set out on a six-month journey to Japan, China, Singapore, Palestine, and Spain. His private reflections on the people he observed on this journey are disturbing in that they imply—and sometimes assert—that other races are biologically inferior to white European observers.[1] Some would say that Einstein here reflects views that were commonplace within his social circles; others would suggest that a "genius" ought to transcend such limits. Still others might suggest that since so many cultural icons turn out to have darker sides, we ought to be cautious about putting anyone on a pedestal in the first place.

A genius, as I see it, isn't someone who is right about everything but someone who is able to open up a new way of seeing things that gives a better quality of vision and a deeper reach than those who went before. And that's what Einstein did. I have tried to present his ideas accurately, fairly, and accessibly. In this final chapter, I want to open up what I consider to be an important conversation, which didn't really get off the ground during Einstein's lifetime.[2] As someone who now writes both as a specialist in the historical and intellectual aspects of science and religion and as a Christian theologian, I want to ask this question: What can those who think seriously about their Christian faith learn from Einstein?

Let me make it clear that I have no intention of forcing Einstein into a Christian (or any other) mold. Einstein clearly wasn't a Christian. He clearly wasn't an atheist, either. He was just Einstein. Yet it is perfectly fair and intellectually legitimate to respect his integrity while at the same time asking what Christians might *learn* from Einstein and how his ideas might feed into wider Christian reflection on a range of important themes. What critical questions does he raise that need to be addressed? What ideas and approaches did he develop that might be useful to Christians in developing their own approaches? Einstein always made it clear that he was not a theologian in any sense of the term. Yet that does not stop theologians from reflecting on his ideas from their own perspectives.

There is a long Christian history of engagement with leading thinkers (particularly philosophers) who are not Christian—not with any intention of falsely claiming that they are really Christians but rather with the intention of developing a dialogue with them. Examples of this include the early Christian engagement with Plato, the medieval exploration of Aristotle's intellectual legacy, and the twentieth-century engagement with philosophers such as Martin Heidegger or Ludwig Wittgenstein. What I propose is to engage Einstein in the same way—respectfully but critically.

Yet there is another point that needs to be made here. Einstein was very reluctant to talk about the "religious" implications of his theories—including his theory of relativity. He took the view that he had clarified the best way of conceiving the ordering of our universe but did not see this as impacting the world of religion. In part, of course, this reflects his own distinct and rather idiosyncratic idea of religion. Yet Einstein "lived close to the frontier between physics and metaphysics."[3] He was intensely alert to the fact that the scientific enterprise rested on certain implicit metaphysical assumptions that could not be proved to be true, such as the reality of a physical world that was open to scientific investigation. It is therefore fair and reasonable to ask, What set of metaphysical assumptions make the most sense of the scientific enterprise? How does a Christian way of seeing the world allow us to both explain the successes of science and identify its limits?

Einstein is an excellent dialogue partner, partly because he is so interesting, but mainly because he is an outstanding scientist with an obvious interest in what the philosopher of science Karl Popper called "ultimate questions." In a lecture at Cambridge University in November 1977, Popper emphasized his own scientific and rational credentials yet distanced himself from those who made "exaggerated claims for science":[4] "I am against intellectual arrogance, and especially against the misconceived claim that we have the truth in our pockets, or that we can approach certainty." As Popper indicated in the lecture, such arrogance or overconfidence is often known as "scientism."

While celebrating the successes of science, Popper warned of the need to recognize and respect its limits: "It is important to realize that science does not make assertions about ultimate questions—about the riddles of existence, or about man's task in this world." Popper insisted that these "ultimate questions" about meaning, purpose, and value could not be properly answered by science. That, however, did not mean they were unanswerable. "The fact that science cannot make any pronouncement about ethical principles has been misinterpreted as indicating that there are no such principles." Such principles, Popper declared, did indeed exist but had to be discovered by means other than the natural sciences. "Some great scientists, and many lesser ones, have misunderstood the situation." I very much doubt if Popper would include Einstein among those who have failed to grasp the point at issue.

Big Pictures—Not a "God of the Gaps"

Christians aim to defend the reasonableness of their faith in many ways, especially in the light of challenges and new questions raised by the development of the natural sciences. One instinct that some Christian apologists follow is to point to the scientifically inexplicable and interpret this as evidence of the need to invoke God to give a coherent account of the universe. The phrase "God of the gaps" is now widely used to describe this approach.

Although others identified and criticized this approach in the first half of the twentieth century, the Oxford theoretical chemist Charles A. Coulson mounted what is now generally seen as the definitive rebuttal of such an approach in his 1956 work *Science and Christian Belief*.[5] Coulson saw this inadequate way of thinking as contracting the area within which God might be known in the first place and impoverishing the intellectual vision of God in the second. This "is a God who leaves Nature still unexplained," while sneaking in "'through the loopholes' of the laws of nature." God was rather to be seen not in gaps that were unexplained but in the grander observation that human beings were able to make so much sense of reality. The capacity of science to explain itself requires explanation—and that means finding a "bigger picture" that makes sense of this observation.

In his lecture on Einstein, given at the Eidgenössischen Technischen Hochschule in Zürich in February 1979, Friedrich Dürrenmatt emphasized Einstein's deep concern for discerning the complexity and coherence of our universe.[6] As we have seen, Einstein is a big-picture thinker, convinced that the role of the natural scientist is to uncover the deeper patterns and structures that account for our observations of our world. A grand theory would give an intellectual or imaginative framework that disclosed *die Einheitlichkeit* of our universe, even if this fundamental unity was displayed in multiple manners. Einstein thus had little interest in explanatory anomalies or gaps in our understanding, except insofar as these might open the way to the emergence of a richer theory (in much the same way as the anomalous behavior of the planet Mercury, which was inexplicable within a Newtonian framework, could be accommodated by the notion of the gravitational warping of space-time).

Christianity as a "Big Picture" of Life

This naturally leads us to ask what forms of Christianity might encourage such "big picture" thinking. It is clear that there are many ways of understanding Christianity that do not see this as an integral aspect of faith. Many forms of the movement known as Pietism, for example, hold that the Christian's responsibility is to focus on a personal devotional life rather than to become preoccupied with intellectual issues. Others would suggest that Christianity is primarily a religion of *salvation* and that a concern with offering an *explanation* of our world does not feature prominently (if it features at all) in the New Testament.

Others, however, will suggest that while Christianity does indeed encourage a "discipleship of the heart," there is also an obligation to develop a "discipleship of the mind." One such writer is the noted Oxford literary critic and writer C. S. Lewis, whose personal journey from atheism to Christianity resulted from his judgment that Christianity offered a better envisioning of reality, coupled with an intellectual capaciousness capable of accommodating the successes of the natural sciences and appreciating their limits.

For Lewis, the Christian faith offers us a means of seeing things properly—as they really are, despite their outward appearances. This theme emerges in his first significant religious work, *The Pilgrim's Regress* (1933), but is developed and expanded throughout his writings. Lewis's commendation of the Christian faith rests partly on his realization that it offers a capacious and deeply satisfying vision of reality—a way of looking at things that simultaneously allows us to both discern its complexity and affirm its interconnectedness.[7]

Other Christian writers have developed similar approaches. For example, the Oxford philosopher of religion Basil Mitchell suggested that Christianity ought to be seen as a worldview or metaphysical system that attempts to make sense of human experience as a whole: "Traditional Christian theism may be regarded as a worldview or metaphysical system which is in competition with other such systems and must be judged by its capacity to make sense of all the available evidence."[8]

A third Christian writer may also be noted here. Dorothy L. Sayers is best remembered for her crime fiction, now seen as one of the highlights of the "Golden Age" of British detective novels during the 1930s. Sayers saw human beings as searching for "patterns" in life and explored this theme in her Peter Wimsey detective novels as well as her religious writings. How can we find the best explanation of what we observe?[9] Sayers realized that the detective novel appeals to our implicit belief in the intrinsic rationality of the world around us and our ability to discover its deeper patterns. Her more explicitly theological works aim to develop this point. Christianity gave her a tool by which she might "make sense of the universe," disclosing hidden patterns and allowing her to see meaning within its otherwise opaque mysteries.

Einstein himself opens up some very important possibilities here. For example, consider his remarkable statement that "the eternal mystery of the world is its comprehensibility."[10] Why should this be so? What bigger picture of our world might be offered that makes sense of our ability to make sense of things? The American psychologist William James remarked that religious belief could be seen as a quest for an "unseen order of some kind in which the riddles of the natural order may be found and explained."[11] One such riddle is that posed by Einstein himself, yet there are others that also claim our attention.

John Polkinghorne, Cambridge quantum physicist turned theologian, is acutely alert to these issues, writing of what he terms "meta-questions," which "arise from our scientific experience and understanding but which point us beyond what science by itself can presume to speak about."[12] So what "meta-questions" does Polkinghorne have in mind? The first he notes is one we have already considered: Why is the physical universe so rationally transparent to us that we can discern its pattern and structures? Why is it that some of the most beautiful patterns proposed by pure mathematicians are actually found to occur in the structure of the physical world? How are we to account for the ability of mathematics to model so accurately the fundamental structures of the universe?[13]

Polkinghorne's response is that Christianity provides an intellectual framework that makes sense of our ability to understand our universe: "Theology can render this discovery intelligible, through its understanding that the Mind of the Creator is the source of the wonderful order of the world."[14] What religion can offer to science is to place scientific insights within a broader and deeper context of intelligibility, providing a framework that is able to hold together both objective and subjective accounts of reality. The parallels with Einstein's own way of thinking here are remarkable.

Yet Polkinghorne builds on this way of thinking that is grounded in the deep structures of the universe, arguing that these structures are best explained by the central themes of the Christian faith. Science "needs to be considered in the wider and deeper context of intelligibility that a belief in God affords."[15] An intellectual bridge can be constructed between Einstein's idea of a "superior mind" disclosed in and through the order of the universe and the more specifically Christian vision of God. Einstein is here the starting point of a journey that leads from an impersonal transcendent order to the Christian idea of a personal God, disclosed both in the created order and especially in and through Christ.

The Two Books: Christianity and the Natural Sciences

Complex relationships are often best envisaged using images or metaphors, which can act as powerful cognitive tools to help us make sense of our world. Such metaphors are helpful in the imaginative representations of disciplinary boundaries, the mapping of complex structures, and the framing of potential relationships. As we noted earlier in this work, perhaps the most influential metaphor used in Western culture to frame the relation of

Christianity and the natural sciences is that of "conflict" or "warfare."[16] Though long discredited by scholarship, and seriously out of place in other cultural contexts (such as India), it retains an appeal that ensures its constant and uncritical repetition in the Western popular media.

Yet there are alternative metaphors, originating within contexts where the potential for productive and meaningful conversations between Christian theology and the natural sciences was acknowledged and actualized. The most important of these is the metaphor of the "two books," which emerged during the Renaissance. One of the clearest statements of this approach is found in the writings of Sir Thomas Browne (1605–1682), particularly his 1643 work *Religio Medici* ("The Religion of a Physician"):

> There are two Books from whence I collect
> my Divinity; besides that written one of God,
> another of His servant Nature, that universal
> and publick Manuscript, that lies expans'd unto
> the Eyes of all: those that never saw Him in the
> one, have discovered Him in the other.[17]

The metaphor of the "two books" was widely used to affirm and preserve the distinctiveness of the natural sciences and Christian theology on the one hand yet to affirm their capacity for interaction on the other. Both, seen from a Christian perspective, originated from God, who is their author or creator. God could thus be known, in different ways and to different extents, through each of these books individually and even more clearly through setting them side by side. Sixteenth-century Reformed confessions of faith—such as the Belgic Confession—affirmed that "the universe is before our eyes like a beautiful book," designed to encourage us to "ponder the invisible things of God," while simultaneously emphasizing that the Bible both clarified and extended this knowledge of God, setting it on a more reliable foundation.

The Metaphor of the Two Books

The metaphor of God's two books is a specifically Christian idea, which reflects specifically Christian ways of thinking. It rests on a fundamental belief that the God who created the world is also the God who is disclosed in and through the Christian Bible. Without this underlying and informing assumption, the "two books" might be seen as two unrelated matters. The link between them is established and safeguarded by the Christian theological assumption of a creator God who is revealed in the Bible. For this reason, it remains both a valid and a valuable conceptual tool for the Christian community as it seeks to frame and engage the natural sciences.

The idea of the "two books" is found in a less developed form in the writings of John Calvin, who recognizes a "natural knowledge of God" accessible to all human beings that is independent of God's revelation in Scripture. Such knowledge can arise from a "sense of divinity" within us or from reflection on the world of nature. For our purposes, we shall focus on this second source of a natural knowledge of God. Calvin then argues that what may be ascertained about God from nature is confirmed, explained, expanded, and enriched in Scripture. Einstein is an important witness to such a "natural knowledge of God." As we have seen, he constantly speaks of a "superior mind" behind the universe and links the intelligibility of the world with this mind or force.

Calvin would invite us to see Einstein's insights as an excellent starting point for our reflections rather than their final destination. For Calvin, human beings possess "weak vision," which prevents us from fully seeing the God to whom nature points.[18] Nature often seems to be fragmentary. As Einstein himself emphasized, we need a *theoria*—a way of seeing our world correctly—that allows us to discern its fundamental unity. The unity of our universe is not something we observe; it is something we believe to be true. For Calvin, that unity can be discerned and affirmed through a close reading of Scripture, which functions as a set of "spectacles coordinating the otherwise confused knowledge of God in our minds."

Calvin thus argues that the Christian faith is able to expand on this natural intuition or perception—evident in Einstein—that there is a mind behind the universe. At one level, it affirms this intuition; at another, it insists that there is more that can be said, including, for example, the human need for redemption and how this hope is able to transform human life. It opens the door to a grander vision of ourselves and our world.

So how does this metaphor of the "two books" help us bring Einstein into the conversation about science and faith? Perhaps the most important point to emphasize is that, from a Christian standpoint, each of these two "books" takes a different form while sharing a common author. To use a common Renaissance idea, the motto of the "book of nature" is *Deus dixit et facta sunt* ("God spoke and things were created"), whereas that of the "book of Scripture" is *Deus dixit et scripta sunt* ("God spoke and things were written down").[19] The idea of complementarity or enrichment is thus integral to the metaphor. It is an image that invites us—perhaps even *requires* us—to ask how these two books can be read together in such a way that a deeper and richer understanding of their author may be achieved. Taken together, they offer us stereoscopic depth, rather than a shallow and superficial reading of our world.

Einstein points us to the book of nature, highlighting its mystery, its elegance and order, and his deep sense of religiosity in its presence. A Christian could then set Einstein's reading of the "book of Nature" alongside her own reading of the "book of Scripture" and find that its features were brought into sharper focus as a result. This is not to suggest that Einstein is a Christian or that he offers a Christian reading of the natural world. The point is that he offers a reading of the natural world that resonates or chimes in with the Christian faith.

For example, think of Einstein's many references to a "mind" behind the universe—as, for example, we see it in his expression of a "firm belief . . . in a superior mind that reveals itself in the world of experience."[20] For Einstein, this notion of a "superior mind" was the essence of his "conception of God." The quantum physicist and theologian John Polkinghorne helps us see how this correlates with a Christian perspective:

> The physical universe seems shot through
> with signs of mind. That is indeed so, says the
> theist, for it is God's Mind that lies behind
> its rational beauty. I do not offer this as a
> knockdown argument for theism—there are
> no such arguments, either for or against—but
> as a satisfying insight which finds a consistent
> place in a theistic view of the world.[21]

Einstein would certainly affirm Polkinghorne's starting point, whatever he might think of his final conclusion.

So what difference does a Christian reading of Einstein allow? Earlier we cited the American psychologist William James in reflecting on Einstein. At this point, we may turn to James again and reflect on one of his more remarkable insights: "At a single stroke, [theism] changes the dead blank *it* of the world into a living *thou*, with whom the whole man may have dealings."[22] What does James mean by this? The basic point he is making is that recognizing God is creator—as we find this idea, for example, in Christianity—means we cannot think of the world as an impersonal object. *It has personal associations.* Let me explain.

Let's reflect on two scenarios. First, imagine an artist painting a beautiful landscape. A painting is not the same as a photograph. The painter puts something of herself into the work of art. In viewing that painting, we do more than see the landscape; we learn something about the artist. In the same way, the creation discloses the Creator. "The heavens declare the glory of God" (Psalm 19:1). God's wisdom and character is thus made known through what God has created. Second, imagine an item that was given to you by someone who is very special to you—perhaps your parents. It may be an impersonal object, but it has deeply personal associations for you. It reminds you of that special person.

That's the point that James is making. Christianity allows us to see the creation as something special, shot through with signs of a person. Einstein himself didn't believe in a personal God. Looking at the universe through a Christian lens, however, allows us to see it as a witness to the greater beauty and wisdom of a loving God. As John Polkinghorne remarks, science discloses a world which in its deep intelligibility could fittingly be described as shot through with signs of mind.[23] Christianity provides a framework that allows us to move from a universe displaying "signs of mind" to a universe that discloses the presence of a loving God.

Conclusion: Einstein and the Bigger Picture

This book has celebrated Einstein's achievements as a scientist and human being while also opening up some deeper questions that hover over the face of the scientific enterprise. As we have seen, the natural sciences deal with the objective aspects of reality. That's essential for the scientific method. Yet we know that there is more to figuring out what life is all about than discovering how things happen. We want to know whether there is meaning, purpose, and value in what is happening. And the scientific method simply does not deal with these "ultimate questions."

Einstein, as we saw earlier, was puzzled and troubled by the failure of scientific objectivity to deal with the subjective aspects of life. At times he found it difficult to understand why the subjective side of life mattered so much to people. When his close friend Michele Besso died in March 1955, Einstein wrote a letter of condolence to his family. Michele, he remarked, had "departed from this strange world a little ahead of me." As a scientist, Einstein was clear that this event was insignificant.

"That means nothing. For believing physicists like us know that the distinction between past, present and future has only the meaning of a persistent illusion."[24] Yet others might reasonably observe that most people think that moving from life to death is *significant*. It's about more than a mere change of position on a set of coordinates!

Yet at other times, he realized these deep, existential questions mattered to people (including scientists!), even if they were "simply beyond the reach of science." That's why we seek for a richer vision of reality that engages both the cognitive and existential dimensions of life. Einstein may not have solved these questions for himself. He is, however, a great dialogue partner in trying to appreciate their importance, and a productive starting point for some serious apologetic conversations.

In closing, it is interesting to reflect on the experiences of Paul Kalanithi (1977–2015), a promising neurosurgeon who died of metastatic lung cancer at the age of thirty-seven before he could ever practice as a fully qualified surgeon. His bestseller *When Breath Becomes Air* was written during his final illness and published posthumously. It is a remarkable testimony to the importance of the ideas we have been considering in this book. Kalanithi loved science and valued its objectivity. However, he found it failed to engage—and could not engage—the deep and urgent questions that really mattered to him and which became increasingly important as his illness progressed. Scientific knowledge, he concluded, was "inapplicable to the existential, visceral nature of human life."[25]

Human beings are driven to find something deeper than what can be found through a superficial examination of the empirical world. The great Old Testament wisdom books—well known to Einstein—speak of wisdom as something that lies beneath the surface of our world. It is something that has to be searched out, not something that stares us in the face. That's what the human quest for meaning is all about.

No reasonable person would ask us to give up the quest for scientific objectivity. Nor would they ask us to give up the quest for meaning and value in our lives. We need to be able to make some response to Kalanithi's ultimate question: What makes life meaningful enough to go on living? Einstein realized that both the objective and subjective realms are important and that they need to be affirmed and held together. Perhaps he himself never found a synthesis that entirely satisfied him, but he certainly provided a road map for those who wish to explore the matter further. A theory of everything that matters engages both our objective and subjective concerns, linking them together in a coherent whole. As the Cambridge physicist Alexander Wood pointed out, "our first demand of religion" is that it "should illumine life and make it a whole."[26]

I do not suggest that Christianity alone provides a way of seeing things that allows us to hold together these objective and subjective worlds; that would be arrogant and inaccurate. Yet I cannot overlook the fact that it *does* hold them together and allows them to be seen as part of a greater whole, rather than as disconnected realms of thought. We need a theory of everything that matters to us if we are to live wisely and meaningfully in this complex world. And maybe Einstein can help us develop and consolidate our own "big pictures" of reality. He's helped me think these things through— and I have not the slightest doubt he can help others as well.

notes

INTRODUCTION: ALBERT EINSTEIN: THE WORLD'S FAVORITE GENIUS

1. Einstein, *Ideas and Opinions*, 45.

CHAPTER 1: APPROACHING EINSTEIN: THE WONDER OF NATURE

1. For the background, see Sponsel, "Constructing a 'Revolution in Science.'"
2. Brian, *Einstein*, 191.
3. Pais, *"Subtle Is the Lord"*, 346.
4. Eddington, *The Mathematical Theory of Relativity.*
5. Overbye, "Gravitational Waves Detected, Confirming Einstein's Theory."
6. Einstein, *Ideas and Opinions*, 38.
7. Ortega y Gasset, "El origen deportivo del estado," 259.
8. Rushdie, *Is Nothing Sacred?*, 8.
9. For Coulson's views, see McGrath, *Enriching Our Vision of Reality*, 27–41.
10. Dewey, *The Quest for Certainty*, 255.
11. See Einstein's preface to Planck, *Where Is Science Going?*, 9.
12. See Menninghaus, "Atoms, Quanta, and Relativity in Aldous Huxley's Analogical Mode of Thinking."
13. Woolf, *The Diary of Virginia Woolf*, vol. 3, 68. This was a common misunderstanding at the time of the "twin paradox," in which someone could age more slowly than someone else by approaching the speed of light.

14. Einstein in conversation with William Hermanns: Hermanns, *Einstein and the Poet*, 132.
15. Letter to Eberhard Zschimmer, dated September 30, 1921.
16. Dawkins, *The God Delusion*, 18.
17. Dawkins uses Max Jammer's excellent work *Einstein and Religion* as a source for his citations for Einstein, perhaps predictably failing to quote from those that speak of God or a transcendent reality behind or beyond creation.
18. Letter to an unidentified recipient, dated August 7, 1941. Einstein Archive, Reel 54-927. For comment, see Jammer, *Einstein and Religion*, 97.
19. Jammer, *Einstein and Religion*, 150. Emphasis mine.

CHAPTER 2: THE OLD WORLD: NEWTON'S CLOCKWORK UNIVERSE

1. For the science, see Kragh, *Conceptions of Cosmos*, 46–65.
2. McKie and de Beer, "Newton's Apple."
3. See Chapman, *England's Leonardo*.
4. Epstein, "Voltaire's Myth of Newton."
5. Newton, *Principia*, 507.
6. See the argument of Curry, "Losing Faith."
7. Sklar, *Space, Time, and Spacetime*, 162.
8. Einstein, "Ernst Mach," 102.
9. Von Soldner, "Ueber die Ablenkung eines Lichtstrals von seiner geradlinigen Bewegung."
10. For a very readable account of what we know about black holes, see Susskind, *The Black Hole War*.
11. Maxwell, *The Scientific Papers*, vol. 2, 244.
12. Michelson, "Some of the Objects and Methods of Physical Science," 15.
13. Newcomb, "The Place of Astronomy among the Sciences," 69–70.
14. Millikan, *Autobiography*, 269–70.
15. See especially Kuhn, *The Structure of Scientific Revolutions*.
16. This "precession of the perihelion" of Mercury takes place at the rate of 574 arcseconds (0.159 degree) per century.
17. Le Verrier, "Théorie du mouvement de Mercure."
18. For the background, see Baum and Sheehan, *In Search of Planet Vulcan*.

CHAPTER 3: A SCIENTIFIC REVOLUTIONARY: EINSTEIN'S FOUR PAPERS OF 1905

1. Iliffe, *Priest of Nature*. For further reflections on the social context of the idea of "genius" at the time, see Fara, *Newton: The Making of Genius*.

2. Albury, "Halley's Ode on the *Principia* of Newton," 27.

3. Letter to John Adams, dated October 28, 1813.

4. For an excellent account of these developments, see Rigden, *Einstein 1905*.

5. For a full account of Einstein's life, career, and achievements, see Isaacson, *Einstein: His Life and Universe*; Pais, *"Subtle Is the Lord."*

6. Brush, "Mach and Atomism."

7. Einstein, "Folgerungen aus den Capillaritätserscheinungen."

8. See, for example, Krstić, *Mileva & Albert Einstein*.

9. For the critical report by the PBS ombudsman into the factual basis of *Einstein's Wife*, see http://www.pbs.org/ombudsman/2006/12/einsteins_wife_the_relative_motion_of_facts.html.

10. The best study of the background to these articles, which deals with these assertions in depth, is Stachel, *Einstein's Miraculous Year*. For a good general account, see Isaacson, *Einstein*, 90–106.

11. See the careful study of Martinez, "Handling Evidence in History."

12. Hertz, "Über den Einfluß des ultravioletten Lichtes auf die electrische Entladung."

13. Lenard, "Über die lichtelektrische Wirkung."

14. Nauenberg, "Max Planck and the Birth of the Quantum Hypothesis."

15. Millikan, "A Direct Photoelectric Determination of Planck's 'h.'"

16. Brown, "A Brief Account of Microscopical Observations on the Particles Contained in the Pollen of Plants."

17. Ford, "Confirming Robert Brown's Observations of Brownian Movement."

18. For what follows, see Maiocchi, "The Case of Brownian Motion."

19. For details, see Maiocchi, "The Case of Brownian Motion," 263–74.

20. See especially Perrin, "Mouvement brownien et réalité moléculaire."

21. Rutherford, "The Scattering of α and β Particles by Matter and the Structure of the Atom."

22. Einstein, "Autobiographische Skizze," 10.

23. Renn, "Einstein as a Disciple of Galileo."

24. I here use material from various sections of Einstein, *Relativity*, although I have presented it in a different order than Einstein for the sake of clarity. This work is very accessible and is recommended as a starting point for serious exploration of both special and general relativity.

25. Einstein, *Relativity*, 17.

26. Davies, *About Time*, 59–65.

27. See, for example, Feuer, "The Social Roots of Einstein's Theory of Relativity," 2.

28. Harman and Thomson, *Moral Relativism and Moral Objectivity*, 3.

29. Sommerfeld, "Philosophie und Physik seit 1900," 99.

30. Feynman, *Six Not-So-Easy Pieces*, 73–74.

31. Russell, "Relativity," 331.

32. Einstein and Infeld, *The Evolution of Physics*, 77.

33. Boughn, "Fritz Hasenöhrl and $E = mc^2$."

34. For the debate, see Ohanian, "Did Einstein Prove $E = mc^2$?"

35. Hecht, "How Einstein Confirmed $E = mc^2$."

36. The best of these, in my view, is his 1935 article "Elementary Derivation of the Equivalence of Mass and Energy."

37. See Duhem, *La science allemande*. This work was written during the First World War, which might help explain its hostility towards German science.

38. Einstein, *Ideas and Opinions*, 226.

39. Wertheimer, *Productive Thinking*, 213–28. For this remarkable relationship, see Miller, "Albert Einstein and Max Wertheimer."

40. Cockcroft and Walton, "Disintegration of Lithium by Swift Protons."

41. Cockcroft and Walton noted that the sum of the rest-masses of the original proton and the lithium nucleus was 8.0176 atomic mass units. However, the sum of the rest-masses of the two alpha particles produced by this reaction was 8.0022 atomic mass units. The reaction thus led to a loss of mass of 0.0154 atomic mass units.

42. Speech given to the British Association for the Advancement of Science, September 11, 1933.

43. See Lanouette and Silard, *Genius in the Shadows*. There is a good account of Einstein's involvement with the development of the atomic bomb in Isaacson, *Einstein*, 471–86.

44. Jerome, *The Einstein File*.

45. Einstein, "On My Participation in the Atom Bomb Project."

46. Einstein, *Essays in Humanism*, 24–25.

CHAPTER 4: THE THEORY OF GENERAL RELATIVITY: FINAL FORMULATION AND CONFIRMATION

1. Technically, Einstein was appointed as "außerordentlicher Professor," which is not easy to convert into an English-language equivalent.
2. Maxwell, "A Dynamical Theory of the Electromagnetic Field," 466.
3. Wheeler, *Geons, Black Holes and Quantum Foam*, 235.
4. Einstein, *Ideas and Opinions*, 100–105.
5. O'Raifeartaigh et al., "One Hundred Years of the Cosmological Constant."
6. Einstein, "Erklärung der Perihelbewegung des Merkur aus der allgemeinen Relativitätstheorie."
7. See Holberg, "Sirius B and the Measurement of the Gravitational Redshift."
8. For an excellent study, see Almassi, "Trust in Expert Testimony."
9. Sponsel, "Constructing a 'Revolution in Science,'" 448.
10. "Lights All Askew."
11. See especially Waller, *Einstein's Luck*, 102–3.
12. Harvey, "Gravitational Deflection of Light."
13. Cited in Holton, "Einstein's Search," 1–15.
14. Pais, *"Subtle Is the Lord"*, 30.
15. Goenner, "The Reaction to Relativity Theory I."
16. These are discussed in van Dongen, "Reactionaries and Einstein's Fame."
17. Thüring, "Physik und Astronomie in jüdischen Händen."
18. For this period of Einstein's life, see Goenner, *Einstein in Berlin 1914–1933*; Hoffmann, *Einstein's Berlin*.
19. See, for example, Hentschel, *Interpretationen*.
20. For what follows, see Friedman, *The Politics of Excellence*.
21. Ravin, "Gullstrand, Einstein, and the Nobel Prize."
22. Rowe and Schulmann, *Einstein on Politics*, 151–52.
23. Forster, "What I Believe," 67.
24. For the story, see Paterniti, *Driving Mr. Albert*.
25. Letter to Willem de Sitter, in *Collected Papers,* vol. 8.
26. This quote is found in a private letter from Einstein to the Hungarian physicist Cornelius Lanczos, who was then based at Princeton, dated March 12, 1942. Einstein wrote, "It seems hard to sneak a look at God's cards. But that he plays dice and uses 'telepathic' methods (as the present quantum theory requires of him) is something that

I cannot believe for a single moment." This is unfortunately often simplified to "God does not play dice." For a good discussion, see Ghirardi, *Sneaking a Look at God's Cards*, 149–64.

27. Paty, "The Nature of Einstein's Objections to the Copenhagen Interpretation of Quantum Mechanics."

28. Van Dongen, *Einstein's Unification*, 186.

29. Greene, *The Elegant Universe*, 15.

30. For a good discussion, see Paty, *Einstein Philosophe*.

CHAPTER 5: EINSTEIN AND THE BIGGER PICTURE: WEAVING THINGS TOGETHER

1. On the importance of this, see Carroll, *The Big Picture*, 69.

2. Letter to Robert Thornton, dated December 7, 1944. Einstein Archive, Reel 6-574.

3. Planck, *Where Is Science Going?*, 218.

4. Planck, *Where Is Science Going?*, 214.

5. Einstein and Infeld, *The Evolution of Physics*, 159.

6. Einstein here anticipates aspects of what is now known as the "unificationist" approach to scientific explanation.

7. Einstein, *Cosmic Religion with Other Opinions and Aphorisms*, 97.

8. Downie, "Science and the Imagination in the Age of Reason"; Rocke, *Image and Reality*.

9. Einstein, *Cosmic Religion with Other Opinions and Aphorisms*, 97.

10. Wertheimer, *Productive Thinking*, 213–28.

11. Einstein, *Mein Weltbild*. Although written in German, this book was published in Amsterdam in 1934 by Querido Verlag, which published titles of German writers in exile from Nazi Germany. Emanuel Querido, who established this publishing house, was killed by the Nazis in Sobibor extermination camp on July 23, 1943.

12. Einstein, *Ideas and Opinions*, 292.

13. Bergmann, "The Quest for Unity."

14. Metz, *Meaning in Life*, 249.

15. Pears, *Hume's System*, 99. See also Lynch, "Hume and the Limits of Reason."

16. Letter to Heinrich Zangger, dated March 10, 1914, in *Collected Papers*, vol. 5, 381.

17. Kessler, *The Diaries of a Cosmopolitan 1918–1937*, 332.

18. Einstein, "Elsbachs Buch," 1685.

19. For example, see Einstein, *Ideas and Opinions*, 224–27.

20. Heisenberg, "Die Kopenhagener Deutung der Quantentheorie," 85. Emphasis mine.

21. Isaacson, *Einstein*, 14.

22. Schilpp, *Albert Einstein: Philosopher-Scientist*, 47.

23. Chaplin, *My Autobiography*, 317. I have found no record of what music Einstein actually played during this process of reflection.

24. For comment, see Moszkowski, *Einstein the Searcher*, 222.

25. Miller, "A Genius Finds Inspiration in the Music of Another."

26. Einstein, *Cosmic Religion with Other Opinions and Aphorisms*, 100.

27. Hawking, *A Brief History of Time*, 193.

28. Rigden, *Einstein 1905*, 147–49.

29. Letter to an unidentified recipient, dated August 7, 1941. Einstein Archive, Reel 54-927.

30. For comment, see Jammer, *Einstein and Religion*, 125–27.

31. See Wilczek, *A Beautiful Question*.

32. Einstein, *Cosmic Religion with Other Opinions and Aphorisms*, 84.

33. Einstein, *Ideas and Opinions*, 151–58.

34. Einstein, *Ideas and Opinions*, 152.

35. For the issue, and its wider implications, see McGrath, *The Territories of Human Reason*.

36. Einstein, *Ideas and Opinions*, 41–49.

37. Einstein, *Ideas and Opinions*, 41–42.

38. Einstein, *Ideas and Opinions*, 148. For comment, see Michalos, "Einstein, Ethics, and Science."

39. Einstein, *Ideas and Opinions*, 42.

40. Rowe and Schulmann, *Einstein on Politics*, xxiv–xxv.

41. For what follows, see the fuller discussion in Midgley, *Science and Poetry*, 170–213.

42. For Midgley's critique of aggressive reductionisms, which insist we use only one map for everything, see Midgley, "Reductive Megalomania."

CHAPTER 6: A "FIRM BELIEF IN A SUPERIOR MIND": EINSTEIN ON RELIGION

1. For the complexity of the term *religion*, see Harrison, "The Pragmatics of Defining Religion."

2. Dürrenmatt, "Albert Einstein," 58.

3. Einstein, *Ideas and Opinions*, 262.

4. Jammer, *Einstein and Religion*, 150.

5. Dawkins, *The God Delusion*, 35.

6. Einstein, *Ideas and Opinions*, 38–39.

7. This is Einstein's famous response in 1929 to Rabbi Herbert S. Goldstein, who asked him whether he believed in God: Jammer, *Einstein and Religion*, 49.

8. For a good summary of Einstein's relation to Spinoza, see Jammer, *Einstein and Religion*, 43–51.

9. Although some trace aspects of Einstein's scientific theories back to the influence of this Dutch Jewish philosopher, these associations need to be treated with caution. See Jammer, *Einstein and Religion*, 46.

10. Michalos, "Einstein, Ethics, and Science," 347–48.

11. The best study remains Jammer, *Einstein and Religion*, which scrupulously documents the multiple elements of his understanding of religion.

12. Dennett, *Breaking the Spell*, 9.

13. Southwold, "Buddhism and the Definition of Religion."

14. My own analysis leads me to more or less the same conclusions as Jammer's.

15. Letter to Eduard Büsching, dated October 25, 1929, quoted in Jammer, *Einstein and Religion*, 51.

16. Einstein, *Ideas and Opinions*, 47–48.

17. Einstein, *Ideas and Opinions*, 38.

18. Letter to Eric Gutkind, dated January 3, 1954.

19. Jammer, *Einstein and Religion*, 50–51. "Karl Eddi" was a pseudonym for the German journalist Eduard Karl Büsching, author of a book titled *Es gibt keinen Gott* ("There is no God"). On reading the book, Einstein suggested that its argument was such that it ought to have a different title: *Es gibt keinen persönlichen Gott* ("There is no personal God").

20. Kessler, *The Diaries of a Cosmopolitan 1918–1937*, 322.

21. This is the view, for example, of Richard Dawkins: Dawkins, *The God Delusion*, 18.

22. Nadler, *Spinoza's Heresy*, 140–41.

23. Jammer, *Einstein and Religion*, 125–27.

24. Dukas and Hoffmann, *Albert Einstein, the Human Side*, 39.

25. Letter to Albert Chapple, dated February 23, 1954. Einstein Archive 59–405.

26. Einstein, *The World as I See It,* 90.

27. Einstein, *Ideas and Opinions*, 50.

28. Einstein's letter of January 3, 1954, to the philosopher Eric Gutkind, which is not included in the material reviewed by Jammer, *Einstein and Religion*, makes this point, which is repeated elsewhere in his published writings.

29. Levitin, "The Experimentalist as Humanist."

30. Einstein, *Ideas and Opinions*, 38–39.

31. Harrison, "Sentiments of Devotion and Experimental Philosophy in Seventeenth-Century England."

32. Other scholars offer different categories. For example, Ian Barbour develops four such options: conflict, independence, dialogue, and integration. See Cantor and Kenny, "Barbour's Fourfold Way."

33. Einstein, *Ideas and Opinions*, 41–42.

34. For a systematic dismantling of this work's arguments and evidence, see Numbers, ed., *Galileo Goes to Jail and Other Myths about Science and Religion.*

35. Brooke, *Science and Religion*, 6.

36. Harrison, "Introduction," 4.

37. Einstein, *Ideas and Opinions*, 41–44.

38. Einstein, *Ideas and Opinions*, 41–42.

39. Einstein, *Ideas and Opinions*, 44–49.

40. Einstein, *Ideas and Opinions*, 44.

41. Einstein, *Ideas and Opinions*, 48–49.

42. Einstein, *Ideas and Opinions*, 45.

43. Gould, "Nonoverlapping Magisteria."

44. Einstein, *Ideas and Opinions*, 45.

45. Einstein, *Ideas and Opinions*, 46.

46. Einstein, *Ideas and Opinions*, 50.

47. Earnshaw, *Existentialism*, 1–25.

48. Planck, *Where Is Science Going?*, 217.

49. For a good discussion, see Brueggemann, *The Land.*

50. Weyl, *Philosophy of Mathematics and Natural Science,* 116.

51. Lewis, *Surprised by Joy*, 197.

52. Eddington, *The Nature of the Physical World*, 68. For a good discussion, see Galison, "Minkowski's Space-Time." Although Einstein developed special relativity without any of the mathematical formalism introduced

by Minkowski, similar existential concerns can be raised about both ways of conceptualizing space and time.

53. Schilpp, *The Philosophy of Rudolf Carnap*, 37–38. I have altered the translation of the final section of this sentence to capture the sense of Carnap's original German: ". . . es gäbe etwas Wesentliches am Jetzt, das einfach außerhalb der Reichweite der Wissenschaft liege."

CHAPTER 7: GOD AND A SCIENTIFIC UNIVERSE: TOWARDS A CHRISTIAN READING OF EINSTEIN

1. Einstein, *The Travel Diaries*.
2. The theologian Thomas F. Torrance is a rare example of someone who took Einstein's ideas seriously and explored their significance for Christian thought, particularly in relation to the doctrine of the Incarnation. See especially Torrance, *Theological Science*.
3. Polkinghorne, "Space, Time, and Causality," 975.
4. This and the following quotes are from Popper, "Natural Selection and the Emergence of Mind," 341–42.
5. On Coulson and his approach, see McGrath, *Enriching Our Vision of Reality*, 27–41.
6. Dürrenmatt, "Albert Einstein," 59.
7. On these themes, see McGrath, *The Intellectual World of C. S. Lewis*. There are interesting parallels between Einstein and Lewis on the role of intuition and imagination in discovering the structures of reality that merit closer examination.
8. Mitchell, *The Justification of Religious Belief*, 99.
9. Kenney, *The Remarkable Case of Dorothy L. Sayers*.
10. Einstein, *Ideas and Opinions*, 292.
11. James, *The Will to Believe*, 51.
12. Polkinghorne, "The New Natural Theology." For Polkinghorne's approach to these questions, see McGrath, *Enriching Our Vision of Reality*, 59–73.
13. Polkinghorne, *Science and Creation*, 20–21.
14. Polkinghorne, *Theology in the Context of Science*, xx.
15. Polkinghorne, *Theology in the Context of Science*, 95.
16. For the best recent studies, see Brooke, *Science and Religion*; Harrison, *Territories of Science and Religion*.
17. Thomas Browne, *Religio Medici*, section 16.

18. John Calvin, *Institutes of the Christian Religion*, I.vi.1. More generally, see Adams, "Calvin's View of Natural Knowledge of God."
19. Palmerino, "The Mathematical Characters of Galileo's Book of Nature," 27–28.
20. Einstein, *Ideas and Opinions*, 262.
21. Polkinghorne, "Cross-Traffic between Science and Theology," 146.
22. James, *The Will to Believe*, 127.
23. See, for example, Polkinghorne, *Science and Creation*, 17–33.
24. Letter dated March 21, 1955; *Albert Einstein—Michele Besso Correspondence, 1903–55*, 537–38.
25. Kalanithi, *When Breath Becomes Air*, 170.
26. Wood, *In Pursuit of Truth*, 102.

works consulted

WORKS BY ALBERT EINSTEIN

Albert Einstein—Michele Besso Correspondence, 1903–55. Edited by
Pierre Speziali. Paris: Hermann, 1972.

"Autobiographische Skizze." In *Helle Zeit—Dunkle Zeit. In Memoriam
Albert Einstein*, edited by Carl Seelig, 9–17. Zürich: Europa Verlag,
1956.

"A Brief Outline of the Development of the Theory of Relativity."
Nature 106 (1921): 782–84.

*The Collected Papers of Albert Einstein, Volume 5: The Swiss Years:
Correspondence, 1902–1914.* Edited by Martin J. Klein, A. J. Kox,
and Robert Schulmann. Princeton, NJ: Princeton University Press,
1993.

*The Collected Papers of Albert Einstein, Volume 8, Part A: The Berlin Years:
Correspondence 1914–1917.* Edited by Robert Schulmann, A. J.
Kox, Michel Janssen, and József Illy. The Digital Einstein Papers,
Princeton University Press. https://einsteinpapers.press.princeton.
edu/vol8a-doc/?startBookmarkIdx=200.

Cosmic Religion with Other Opinions and Aphorisms. New York: Covici-
Friede Inc., 1931.

"Die formale Grundlage der allgemeinen Relativitätstheorie."
*Sitzungsberichte der Königlich Preussischen Akademie der Wissenschaften
zu Berlin* (1914): 1030–85.

"Die Grundlage der allgemeinen Relativitätstheorie." *Annalen der Physik*
354, no. 7 (1916): 769–822.

Eine neue Bestimmung der Moleküldimensionen. Bern: Buchdruckerei
K. J. Wyss, 1906.

"Elementary Derivation of the Equivalence of Mass and Energy."
   Bulletin of the American Mathematical Society 41 (1935): 223–30.
"Elsbachs Buch: Kant und Einstein." *Deutsche Literaturzeitung für Kritik
   der internationalen Wissenschaft* 45 (1924): 1685–92.
"Erklärung der Perihelbewegung des Merkur aus der allgemeinen
   Relativitätstheorie." *Sitzungsberichte der Königlich Preußischen
   Akademie der Wissenschaften* 47 (1915): 831–39.
"Ernst Mach." *Physikalische Zeitschrift* 7 (1916): 101–4.
Essays in Humanism. New York: Philosophical Library, 1950.
"Folgerungen aus den Capillaritätserscheinungen." *Annalen der Physik* 4
   (1901): 513–23.
Ideas and Opinions. New York: Crown Publishers, 1954.
"Ist die Trägheit eines Körpers von seinem Energieinhalt abhängig?"
   Annalen der Physik 18 (1905): 639.
Mein Weltbild. Amsterdam: Querido Verlag, 1934.
"On My Participation in the Atom Bomb Project." Atomic Archive. http://
   www.atomicarchive.com/Docs/Hiroshima/EinsteinResponse.shtml.
"On the Method of Theoretical Physics." *Philosophy of Science* 1, no. 2
   (1934): 163–69.
"Quanten-Mechanik und Wirklichkeit." *Dialectica* 2 (1948): 320–24.
Relativity: The Special and the General Theory. New York: Barnes and
   Noble, 2004.
"Remarks to the Essays Appearing in This Collective Volume." In *Albert
   Einstein, Philosopher-Scientist,* edited by Paul Arthur Schilpp,
   663–88. Chicago: Open Court, 1970.
*The Travel Diaries of Albert Einstein: The Far East, Palestine, and Spain,
   1922–1923.* Princeton, NJ: Princeton University Press, 2018.
"Über die von der molekularkinetischen Theorie der Wärme geforderte
   Bewegung von in ruhenden Flüssigkeiten suspendierten Teilchen."
   Annalen der Physik 17, no. 8 (1905): 549–60.
"Über einen die Erzeugung und Verwandlung des Lichtes betreffenden
   heuristischen Gesichtspunkt." *Annalen der Physik* 17 (1905): 132–48.
The World as I See It. New York: Covici-Friede Inc., 1934.
"Zur Elektrodynamik bewegter Körper." *Annalen der Physik* 17 (1905):
   891–921.
Einstein, Albert, and Marcel Grossmann. "Entwurf einer verallgemeinerten
   Relativitätstheorie und einer Theorie der Gravitation." *Zeitschrift für
   Mathematik und Physik* 62 (1913): 225–61.

Einstein, Albert, and Leopold Infeld. *The Evolution of Physics: The Growth of Ideas from Early Concepts to Relativity and Quanta*. New York: Simon and Schuster, 1938.

Einstein, Albert, B. Podolsky, and N. Rosen. "Can Quantum-Mechanical Description of Physical Reality Be Considered Complete?" *Physical Review* 47 (1935): 777–80.

WORKS ABOUT EINSTEIN

Bergmann, Peter G. "The Quest for Unity: General Relativity and Unitary Field Theories." *Syracuse Scholar* 1, no. 1 (1979): 9–18.

Bernstein, Jeremy. *Einstein*. New York: Viking Press, 1973.

Bjerknes, Christopher Jon. *Albert Einstein: The Incorrigible Plagiarist*. Downers Grove, IL: XTX, 2002.

Brian, Denis. *Einstein: A Life*. New York: Wiley, 1996.

Canales, Jimena. *The Physicist and the Philosopher: Einstein, Bergson, and the Debate That Changed Our Understanding of Time*. Princeton, NJ: Princeton University Press, 2015.

Clark, Ronald W. *Einstein: The Life and Times*. New York: Avon Books, 1984.

Crease, Robert P., and Alfred S. Goldhaber. *The Quantum Moment: How Planck, Bohr, Einstein, and Heisenberg Taught Us to Love Uncertainty*. New York: W. W. Norton & Company, 2014.

Davies, Paul C. W. *About Time: Einstein's Unfinished Revolution*. New York: Simon & Schuster, 2005.

Dukas, Helen, and Banesh Hoffmann, eds. *Albert Einstein, the Human Side: New Glimpses from His Archives*. Princeton, NJ: Princeton University Press, 1979.

Dürrenmatt, Friedrich. "Albert Einstein." *Naturforschende Gesellschaft in Zürich* 124, no. 8 (1979): 58–73.

Eddington, Arthur Stanley. "Einstein's Theory of Space and Time." *Contemporary Review* 116 (1919): 639–43.

Eddington, Arthur Stanley. *The Mathematical Theory of Relativity*. Cambridge: Cambridge University Press, 1923.

Engler, Gideon. "Einstein and the Most Beautiful Theories in Physics." *International Studies in the Philosophy of Science* 16, no. 1 (2002): 27–37.

Feuer, Lewis S. "The Social Roots of Einstein's Theory of Relativity: Part 1." *Annals of Science* 27, no. 3 (1971): 277–98.

Feuer, Lewis S. "The Social Roots of Einstein's Theory of Relativity: Part 2." *Annals of Science* 27, no. 4 (1971): 313–44.

Feynman, Richard P. *Six Not-So-Easy Pieces: Einstein's Relativity, Symmetry, and Space-Time*. Reading, MA: Addison-Wesley, 1997.

Fölsing, Albrecht. *Albert Einstein: A Biography*. New York: Viking Books, 1997.

Frank, Philipp. *Einstein: His Life and Times*. New York: Knopf, 1947.

Galison, Peter, Gerald James Holton, and Silvan S. Schweber, eds. *Einstein for the 21st Century: His Legacy in Science, Art, and Modern Culture*. Princeton, NJ: Princeton University Press, 2008.

Goenner, Hubert. *Einstein in Berlin 1914–1933*. Munich: Beck, 2005.

Goenner, Hubert. "The Reaction to Relativity Theory I: The Anti-Einstein Campaign in Germany in 1920." *Science in Context* 6 (1993), 107–33.

Gribbin, John, and Mary Gribbin. *Einstein's Masterwork: 1915 and the General Theory of Relativity*. London: Icon, 2015.

Hecht, Eugene. "How Einstein Confirmed $E = mc^2$." *American Journal of Physics* 79, no. 6 (2011). DOI 10.1119/1.3549223.

Hentschel, Klaus. *Interpretationen und Fehlinterpretationen der speziellen und der allgemeinen Relativitätstheorie durch Zeitgenossen Albert Einsteins*. Berlin: Birkhäuser Verlag, 1990.

Hermanns, William. *Einstein and the Poet: In Search of the Cosmic Man*. Brookline Village, MA: Branden Press, 1983.

Hillman, Bruce J., Birgit Ertl-Wagner, and Bernd C. Wagner. *The Man Who Stalked Einstein: How Nazi Scientist Philipp Lenard Changed the Course of History*. Guilford: Lyons Press, 2015.

Hoffmann, Dieter. *Einstein's Berlin: In the Footsteps of a Genius*. Baltimore: Johns Hopkins University Press, 2013.

Holberg, J. B. "Sirius B and the Measurement of the Gravitational Redshift." *Journal for the History of Astronomy* 41, no. 1 (February 2010): 41–64.

Holton, Gerald James, and Yehuda Elkana, eds. *Albert Einstein, Historical and Cultural Perspectives: The Centennial Symposium in Jerusalem*. Princeton, NJ: Princeton University Press, 2014.

Isaacson, Walter. *Einstein: His Life and Universe*. New York: Simon & Schuster, 2007.

Jammer, Max. *Einstein and Religion: Physics and Theology*. Princeton, NJ: Princeton University Press, 1999.

Jerome, Fred. *The Einstein File: J. Edgar Hoover's Secret War against the World's Most Famous Scientist.* New York: Saint Martin's Press, 2002.

Krstić, Djordje. *Mileva & Albert Einstein: ljubezen in skupno znanstveno delo.* Radovljica, Solovenia: Didakta, 2002.

Kumar, Manjit. *Quantum: Einstein, Bohr and the Great Debate about the Nature of Reality.* London: Icon, 2008.

Latour, Bruno. "A Relativistic Account of Einstein's Relativity." *Social Studies of Science* 18 (1988): 3–44.

Martinez, Alberto. "Handling Evidence in History: The Case of Einstein's Wife." *School Science Review* 86, no. 316 (2005): 49–56.

Michalos, Alex C. "Einstein, Ethics, and Science." *Journal of Academic Ethics* 2, no. 4 (2004): 339–54.

Miller, Arthur I. "Albert Einstein and Max Wertheimer: A Gestalt Psychologist's View of the Genesis of Special Relativity Theory." *History of Science* 13 (1975): 75–103.

Miller, Arthur I. "A Genius Finds Inspiration in the Music of Another." *New York Times*, January 31, 2006.

Moszkowski, Alexander. *Einstein the Searcher: His Work Explained from Dialogues with Einstein.* London: Methuen & Co., 1921.

Ohanian, Hans C. "Did Einstein Prove $E = mc^2$?" *Studies in History and Philosophy of Modern Physics* 40 (2009): 167–73.

O'Raifeartaigh, Cormac, Michael O'Keeffe, Werner Nahm, and Simon Mitton. "One Hundred Years of the Cosmological Constant: From 'Superfluous Stunt' to Dark Energy." *European Physical Journal H* 43, no. 1 (2018): 73–117.

Pais, Abraham. *"Subtle Is the Lord": The Science and the Life of Albert Einstein.* Oxford: Oxford University Press, 2005.

Paty, Michel. "Einstein and Spinoza." In *Spinoza and the Sciences*, edited by Marjorie G. Grene and Debra Nails, 267–302. Dordrecht: Kluwer Academic Publishers, 1986.

Paty, Michel. *Einstein Philosophe: La physique comme practique philosophique.* Paris: Presses Universitaires de France, 1993.

Paty, Michel. "The Nature of Einstein's Objections to the Copenhagen Interpretation of Quantum Mechanics." *Foundations of Physics* 25, no. 1 (1995): 183–204.

Popović, Milan, ed. *In Albert's Shadow: The Life and Letters of Mileva Marić, Einstein's First Wife.* Baltimore: Johns Hopkins University Press, 2003.

Ravin, James G. "Gullstrand, Einstein, and the Nobel Prize." *Archives of Ophthalmolology* 117, no. 5 (1999): 670–72.

Renn, Jürgen. "Einstein as a Disciple of Galileo: A Comparative Study of Concept Development in Physics." *Science in Context* 6, no. 1 (1993): 311–41.

Rigden, John S. *Einstein 1905: The Standard of Greatness.* Cambridge, MA: Harvard University Press, 2005.

Rowe, David E., and Robert Schulmann. *Einstein on Politics: His Private Thoughts and Public Stands on Nationalism, Zionism, War, Peace, and the Bomb.* Princeton, NJ: Princeton University Press, 2007.

Russell, Bertrand. "Relativity: Philosophical Consequences." In *Encyclopedia Britannica.* 32 vols. London: Encyclopaedia Britannica, 1926, vol. 31, 331–32.

Sauer, Tilman. "Einstein's Unified Field Theory Program." In *The Cambridge Companion to Einstein,* edited by Michel Janssen and Christoph Lehner, 281–305. Cambridge: Cambridge University Press, 2014.

Schilpp, Paul Arthur. *Albert Einstein: Philosopher-Scientist.* Evanston, IL: Open Court, 1970.

Schönbeck, Charlotte. *Albert Einstein und Philipp Lenard: Antipoden im Spannungsfeld von Physik und Zeitgeschichte.* Berlin: Springer, 2000.

Stachel, John. *Einstein's Miraculous Year: Five Papers That Changed the Face of Physics.* Princeton, NJ: Princeton University Press, 2005.

Weinert, Friedel. "Einstein, Science and Philosophy." *Philosophia Scientiae* 13, no. 1 (2009): 99–133.

OTHER WORKS RELEVANT TO THIS DISCUSSION

Adams, Edward. "Calvin's View of Natural Knowledge of God." *International Journal of Systematic Theology* 3, no. 3 (2001): 280–92.

Albury, W. R. "Halley's Ode on the *Principia* of Newton and the Epicurean Revival in England." *Journal of the History of Ideas* 39, no. 1 (1978): 24–43.

Almassi, Ben. "Trust in Expert Testimony: Eddington's 1919 Eclipse Expedition and the British Response to General Relativity." *History and Philosophy of Modern Physics* 40, no. 1 (2009): 57–67.

Ball, Philip. *Serving the Reich: The Struggle for the Soul of Physics under Hitler.* London: Bodley Head, 2013.

Bamford, Greg. "Popper and His Commentators on the Discovery of Neptune: A Close Shave for the Law of Gravitation?" *Studies in History and Philosophy of Science Part A* 27, no. 2 (1996): 207–32.

Bartelborth, Thomas. "Explanatory Unification." *Synthese* 130 (2002): 91–108.

Baum, Richard, and William Sheehan. *In Search of Planet Vulcan: The Ghost in Newton's Clockwork Universe.* New York: Plenum Press, 1997.

Belkind, Ori. "Newton's Conceptual Argument for Absolute Space." *International Studies in the Philosophy of Science* 21, no. 3 (2007): 271–93.

Beller, Mara. "The Birth of Bohr's Complementarity: The Context and the Dialogues." *Studies in History and Philosophy of Science Part A* 23, no. 1 (1992): 147–80.

Beller, Mara. "Einstein and Bohr's Rhetoric of Complementarity." *Science in Context* 6 , no. 1 (1993): 241–55.

Boughn, Stephen P. "Fritz Hasenöhrl and $E = mc^2$." *European Physical Journal H* 38, no. 2 (2013): 261–78.

Brooke, John Hedley. *Science and Religion: Some Historical Perspectives.* Cambridge: Cambridge University Press, 1991.

Brown, Robert. "A Brief Account of Microscopical Observations on the Particles Contained in the Pollen of Plants." *Edinburgh New Philosophical Journal* (1828): 358–71.

Brueggemann, Walter. *The Land: Place as Gift, Promise, and Challenge in Biblical Faith.* 2nd ed. Philadelphia: Fortress Press, 2002.

Brush, S. G. "Mach and Atomism." *Synthese* 18, no. 2/3 (1968): 192–215.

Cantor, Geoffrey, and Chris Kenny. "Barbour's Fourfold Way: Problems with His Taxonomy of Science-Religion Relationships." *Zygon* 36, no. 4 (December 2001): 765–81.

Carroll, Sean. *The Big Picture: On the Origins of Life, Meaning and the Universe Itself.* London: Oneworld, 2016.

Chaplin, Charlie. *My Autobiography.* London: Penguin Books, 2003.

Chapman, Allan. *England's Leonardo: Robert Hooke and the Seventeenth-Century Scientific Revolution.* Bristol: Institute of Physics Publishing, 2005.

Clarke, Imogen. "How to Manage a Revolution: Isaac Newton in the Early Twentieth Century." *Notes and Records of the Royal Society of London* 68, no. 4 (2014): 323–37.

Cockcroft, J. D., and E. T. S. Walton. "Disintegration of Lithium by Swift Protons." *Nature* 129 (1932): 649.

Coulson, C. A. *Science and Christian Belief.* London: Oxford University Press, 1955.

Curry, Michael F. "Losing Faith: Rationalizing Religion in Early Modern England." *Intersections* 11, no. 2 (2010): 207–41.

Dawkins, Richard. *The God Delusion.* Boston: Houghton Mifflin, 2006.

Dennett, Daniel C. *Breaking the Spell: Religion as a Natural Phenomenon.* New York: Viking Penguin, 2006.

Dewey, John. *The Quest for Certainty: A Study of the Relation of Knowledge and Action.* New York: Capricorn Books, 1960.

Downie, Robin. "Science and the Imagination in the Age of Reason." *Medical Humanities* 27 (2001): 58–63.

Duhem, Pierre. *La science allemande.* Paris: Hermann, 1915.

Earnshaw, Steven. *Existentialism: A Guide for the Perplexed.* London: Continuum, 2006.

Eddington, Arthur Stanley. *The Nature of the Physical World.* Cambridge: Cambridge University Press, 1928.

Eddington, Arthur Stanley. "The Total Eclipse of 1919 May 29 and the Influence of Gravitation on Light." *The Observatory* 42 (1919): 119–22.

Epstein, Julia L. "Voltaire's Myth of Newton." *Pacific Coast Philology* 14, no. 1 (1979): 27–33.

Fara, Patricia. *Newton: The Making of Genius.* New York: Columbia University Press, 2002.

Feynman, Richard P. *The Character of Physical Law.* Boston: MIT Press, 1988.

Ford, Brian J. "Confirming Robert Brown's Observations of Brownian Movement." *Proceedings of the Royal Microscopical Society* 31 (1996): 316–21.

Forster, E. M. "What I Believe." In *Two Cheers for Democracy,* 67–76. San Diego: Harcourt Brace, 1951.

Friedman, Robert Marc. *The Politics of Excellence: Behind the Nobel Prize in Science.* London: W. H. Freeman & Co., 2001.

Galison, Peter L. "Minkowski's Space-Time: From Visual Thinking to the Absolute World." *Historical Studies in the Physical Sciences* 10 (1979): 85–121.

Galton, Francis. *Hereditary Genius: An Inquiry into Its Laws and Consequences*. London: Macmillan, 1869.

Ghirardi, Giancarlo. *Sneaking a Look at God's Cards: Unraveling the Mysteries of Quantum Mechanics*. Translated by Gerald Malsbary. Princeton, NJ: Princeton University Press, 2005.

Gould, Stephen Jay. "Nonoverlapping Magisteria." *Natural History* 106, no. 2 (1997): 16–22.

Greene, Brian. *The Elegant Universe: Superstrings, Hidden Dimensions, and the Quest for the Ultimate Theory*. New York: W. W. Norton, 1999.

Gribbin, John, and Mary Gribbin. *Out of the Shadow of a Giant: Hooke, Halley and the Birth of British Science*. London: Collins, 2016.

Harman, Gilbert, and J. J. Thomson. *Moral Relativism and Moral Objectivity*. Oxford: Blackwell, 1996.

Harrison, Peter. "Introduction." In *The Cambridge Companion to Science and Religion*, edited by Peter Harrison, 1–18. Cambridge: Cambridge University Press, 2010.

Harrison, Peter. "Sentiments of Devotion and Experimental Philosophy in Seventeenth-Century England." *Journal of Medieval and Early Modern Studies* 44, no. 1 (2014): 113–33.

Harrison, Peter. *Territories of Science and Religion*. Chicago: University of Chicago Press, 2015.

Harrison, Victoria. "The Pragmatics of Defining Religion in a Multi-Cultural World." *International Journal for Philosophy of Religion* 59, no. 3 (June 2006): 133–52.

Harvey, Geoffrey M. "Gravitational Deflection of Light: A Re-examination of the Observations of the Solar Eclipse of 1919." *The Observatory* 99 (1979): 195–98.

Hawking, Stephen. *A Brief History of Time: From the Big Bang to Black Holes*. New York: Bantam Books, 1988.

Heisenberg, Werner. "Die Kopenhagener Deutung der Quantentheorie." In *Physik und Philosophie*, 67–85. Stuttgart: Hirzel, 2007.

Hertz, Heinrich. "Über den Einfluβ des ultravioletten Lichtes auf die electrische Entladung." *Annalen der Physik* 267, no. 8 (1887): 983–1000.

Holton, Gerald James. "Einstein's Search for the *Weltbild*." *Proceedings of the American Philosophical Society* 125 (1981): 1–15.

Holton, Gerald James. *Science and Anti-Science*. Cambridge, MA: Harvard University Press, 1993.

Iliffe, Rob. *Priest of Nature: The Religious Worlds of Isaac Newton*. Oxford: Oxford University Press, 2017.

James, William. *The Will to Believe, and Other Essays in Popular Philosophy*. New York: Dover Publications, 1956.

Kalanithi, Paul. *When Breath Becomes Air*. London: Vintage Books, 2017.

Kenney, Catherine McGehee. *The Remarkable Case of Dorothy L. Sayers*. Kent, OH: Kent State University Press, 1990.

Kershaw, Ian. *Hitler: A Biography*. New York: W. W. Norton & Co., 2008.

Kessler, Harry. *The Diaries of a Cosmopolitan 1918–1937*. London: Phoenix, 2000.

Köhne, Julia Barbara. "The Cult of the Genius in Germany and Austria at the Dawn of the Twentieth Century." In *Genealogies of Genius*, edited by Joyce E. Chaplin and Darrin M. McMahon, 115–34. Basingstoke: Palgrave Macmillan, 2016.

Kragh, Helge. *Conceptions of Cosmos from Myths to the Accelerating Universe: A History of Cosmology*. Oxford: Oxford University Press, 2007.

Kuhn, Thomas S. *The Structure of Scientific Revolutions*. 2nd ed. Chicago: University of Chicago Press, 1970.

Lanouette, William, and Bela A. Silard. *Genius in the Shadows: A Biography of Leo Szilard, The Man Behind the Bomb*. New York: Charles Scribner's Sons, 1992.

Lenard, Philipp. "Über die lichtelektrische Wirkung." *Annalen der Physik* 8 (1902): 149–98.

Le Verrier, Urbain J. "Théorie du mouvement de Mercure." *Annales de l'Observatoire Impérial de Paris* 5 (1859): 1–196.

Levitin, Dmitri. "The Experimentalist as Humanist: Robert Boyle on the History of Philosophy." *Annals of Science* 71, no. 2 (2014): 149–82.

Lewis, C. S. *Surprised by Joy*. London: HarperCollins, 2002.

"Lights All Askew in the Heavens." *New York Times*. November 10, 1919. https://www.nytimes.com/1919/11/10/archives/lights-all -askew-in-the-heavens-men-of-science-more-or-less-agog.html.

Lynch, Michael P. "Hume and the Limits of Reason." *Hume Studies* 22, no. 1 (April 1996): 89–104.

Maiocchi, Roberto. "The Case of Brownian Motion." *British Journal for the History of Science* 23, no. 3 (1990): 257–83.

Markley, Robert. "Representing Order: Natural Philosophy, Mathematics, and Theology in the Newtonian Revolution." In *Chaos and Order: Complex Dynamics in Literature and Science*, edited by N. Katherine Hayles, 125–48. Chicago: University of Chicago Press, 1991.

Maxwell, James Clerk. "A Dynamical Theory of the Electromagnetic Field." *Philosophical Transactions of the Royal Society of London* 155 (1865): 459–512.

Maxwell, James Clerk. *The Scientific Papers of James Clerk Maxwell.* 2 vols. Cambridge: Cambridge University Press, 1890.

McGrath, Alister E. *Enriching Our Vision of Reality: Theology and the Natural Sciences in Dialogue*. West Conshohocken, PA: Templeton Foundation, 2016.

McGrath, Alister E. *The Intellectual World of C. S. Lewis.* Oxford: Wiley-Blackwell, 2013.

McGrath, Alister E. *The Territories of Human Reason: Science and Theology in an Age of Multiple Rationalities*. Oxford: Oxford University Press, 2019.

McKie, D., and Gavin R. de Beer. "Newton's Apple." *Notes and Records of the Royal Society of London* 9, no. 1 (1951): 46–54.

McMahon, Darrin M. *Divine Fury: A History of Genius.* New York: Basic Books, 2013.

Menninghaus, Sabine. "Atoms, Quanta, and Relativity in Aldous Huxley's Analogical Mode of Thinking." In *The Perennial Satirist: Essays in Honour of Bernfried Nugel*, edited by Peter Edgerly Firchow, Hermann Josef Real, and Bernfried Nugel, 245–64. Münster: LIT Verlag, 2005.

Metz, Thaddeus. *Meaning in Life: An Analytic Study*. Oxford: Oxford University Press, 2013.

Michelson, Albert A. "Some of the Objects and Methods of Physical Science." *The Quarterly Calendar of the University of Chicago*, no. 3, (August 1894): 14–15.

Midgley, Mary. "Reductive Megalomania." In *Nature's Imagination: The Frontiers of Scientific Vision*, edited by John Cornwell, 133–47. Oxford: Oxford University Press, 1995.

Midgley, Mary. *Science and Poetry.* London: Routledge, 2001.

Millikan, Robert A. *Autobiography*. New York: Prentice-Hall, 1950.

Millikan, Robert A. "A Direct Photoelectric Determination of Planck's 'h'." *Physical Review* 7, no. 3 (1916): 355–88.

Mitchell, Basil. *The Justification of Religious Belief.* London: Macmillan, 1973.

Morrison, Margaret. "One Phenomenon, Many Models: Inconsistency and Complementarity." *Studies in History and Philosophy of Science Part A* 42, no. 2 (2011): 342–51.

Nadler, Steven M. *Spinoza's Heresy: Immortality and the Jewish Mind.* Oxford: Clarendon Press, 2001.

Nauenberg, Michael. "Max Planck and the Birth of the Quantum Hypothesis." *American Journal of Physics* 84, no. 9 (2016): 709–20.

Nersessian, Nancy J. "In the Theoretician's Laboratory: Thought Experimenting as Mental Modeling." *PSA: Proceedings of the Biennial Meeting of the Philosophy of Science Association 1992* 2 (1992): 291–301.

Newcomb, Simon. "The Place of Astronomy among the Sciences." *Sidereal Messenger* 7 (1888): 65–73.

Newton, Sir Isaac. *Newton's Principia: The Mathematical Principles of Natural Philosophy.* New York: Adee, 1846.

Numbers, Ronald L., ed. *Galileo Goes to Jail and Other Myths about Science and Religion.* Cambridge, MA: Harvard University Press, 2009.

Ortega y Gasset, José. "El origen deportivo del estado." *Citius, Altius, Fortius* 9, no. 1–4 (1967): 259–76.

Overbye, Dennis. "Gravitational Waves Detected, Confirming Einstein's Theory." *New York Times*, February 11, 2016. https://www.nytimes.com/2016/02/12/science/ligo-gravitational-waves-black-holes-einstein.html.

Palmerino, Carla Rita. "The Mathematical Characters of Galileo's Book of Nature." In *The Book of Nature in Early Modern and Modern History*, edited by Klaas van Berkel and Arjo Vanderjagt, 27–44. Leuven: Peeters, 2006.

Paterniti, Michael. *Driving Mr. Albert: A Trip across America with Einstein's Brain.* London: Abacus, 2002.

Pears, David. *Hume's System: An Examination of the First Book of His Treatise.* Oxford: Oxford University Press, 1990.

Penrose, Roger. *The Road to Reality: A Complete Guide to the Laws of the Universe.* London: Jonathan Cape, 2004.

Perrin, Jean. "Mouvement brownien et réalité moléculaire." *Annales de chimie et de physique* 18, no. 8 (1909): 5–114.

Planck, Max. *The Universe in the Light of Modern Physics.* London: George Allen & Unwin, 1931.

Planck, Max. *Where Is Science Going?* Translated by James Vincent Murphy. New York: W. W. Norton & Co., 1932.

Polkinghorne, John. "Cross-Traffic between Science and Theology." *Perspectives on Science and Christian Faith* 43, no. 3 (1991): 144–51.

Polkinghorne, John. "The New Natural Theology." *Studies in World Christianity* 1, no. 1 (1995): 41–50.

Polkinghorne, John. *Science and Creation: The Search for Understanding.* London: SPCK, 1988.

Polkinghorne, John. "Space, Time, and Causality." *Zygon* 41, no. 4 (2006): 975–84.

Polkinghorne, John. *Theology in the Context of Science.* New Haven, CT: Yale University Press, 2009.

Popper, Karl. "Natural Selection and the Emergence of Mind." *Dialectica* 32, nos. 3–4 (1978): 339–55.

Rocke, Alan J. *Image and Reality: Kekulé, Kopp, and the Scientific Imagination.* Chicago: University of Chicago Press, 2010.

Rushdie, Salman. "Is Nothing Sacred?" Herbert Read Memorial Lecture, February 6, 1990.

Rutherford, Ernest. "The Scattering of α and β Particles by Matter and the Structure of the Atom." *Philosophical Magazine* 6, no. 21 (1911): 669–88.

Salmon, Wesley C. *Four Decades of Scientific Explanation.* Minneapolis: University of Minnesota Press, 1989.

Schilpp, P. A., ed. *The Philosophy of Rudolf Carnap.* La Salle, IL: Open Court Publishing, 1963.

Schliesser, Eric. "Newton's Challenge to Philosophy." *HOPOS: The Journal of the International Society for the History of Philosophy of Science* 1 (2011): 101–28.

Shaw, George Bernard, and Fred D. Crawford. "Toast to Albert Einstein." *Shaw* 15 (1995): 231–41.

Sklar, Lawrence. *Space, Time, and Spacetime.* Berkeley, CA: University of California Press, 1977.

Snobelen, Stephen D. "The Myth of the Clockwork Universe: Newton, Newtonianism, and the Enlightenment." In *The Persistence of the Sacred in Modern Thought*, edited by Chris L. Firestone and Nathan Jacobs, 149–84. South Bend, IN: University of Notre Dame Press, 2012.

Sommerfeld, Arnold. "Philosophie und Physik seit 1900." *Naturwissenschaftliche Rundschau* I (1948): 97–100.

Southwold, Martin. "Buddhism and the Definition of Religion." *Man: New Series* 13, no. 3 (September 1978): 362–79.

Sponsel, Alistair. "Constructing a 'Revolution in Science': The Campaign to Promote a Favourable Reception for the 1919 Solar Eclipse Experiments." *British Journal for the History of Science* 35, no. 4 (2002): 439–67.

Stachel, John, and Roger Penrose, eds. *Einstein's Miraculous Year: Five Papers That Changed the Face of Physics*. Princeton, NJ: Princeton University Press, 2005.

Stein, Howard. "Newtonian Space-Time." In *The Annus Mirabilis of Sir Isaac Newton 1666–1966*, edited by Robert Palter, 174–200. Cambridge, MA: MIT Press, 1967.

Susskind, Leonard. *The Black Hole War: My Battle with Stephen Hawking to Make the World Safe for Quantum Mechanics*. New York: Back Bay Books, 2009.

Thüring, Bruno. "Physik und Astronomie in jüdischen Händen." *Zeitschrift für die gesamte Naturwissenschafte* 3 (May/June 1937): 55–70.

Torrance, Thomas F. *Theological Science*. London: Oxford University Press, 1969.

van Dongen, Jeroen. *Einstein's Unification*. Cambridge: Cambridge University Press, 2010.

van Dongen, Jeroen. "Reactionaries and Einstein's Fame: 'German Scientists for the Preservation of Pure Science,' Relativity, and the Bad Nauheim Meeting." *Physics in Perspective* 9, no. 2 (June 2007): 212–30.

von Soldner, Johann Georg. "Ueber die Ablenkung eines Lichtstrals von seiner geradlinigen Bewegung, durch die Attraktion eines Weltkörpers, an welchem er nahe vorbegeht." *Berliner Astronomisches Jahrbuch* (1804): 161–72.

Waller, John. *Einstein's Luck: The Truth behind Some of the Greatest Scientific Discoveries*. Oxford: Oxford University Press, 2002.

Weiner, Eric. *The Geography of Genius: A Search for the World's Most Creative Places from Ancient Athens to Silicon Valley.* New York: Simon & Schuster Paperbacks, 2016.

Wertheimer, Max. *Productive Thinking.* New York: Harper & Row, 1959.

Weyl, Hermann. *Philosophy of Mathematics and Natural Science.* Princeton, NJ: Princeton University Press, 1949.

Wheeler, John Archibald. *Geons, Black Holes and Quantum Foam: A Life in Physics.* New York: W. W. Norton & Company, 1998.

White, Andrew Dickson. *A History of the Warfare of Science with Theology in Christendom.* 2 vols. New York: Appleton, 1896.

Wilczek, Frank. *A Beautiful Question: Finding Nature's Deep Design.* London: Penguin Books, 2016.

Wood, Alexander. *In Pursuit of Truth: A Comparative Study in Science and Religion.* London: Student Christian Movement, 1927.

Woolf, Virginia. *The Diary of Virginia Woolf.* Edited by Anne Olivier Bell and Andrew McNeillie. 5 vols. Harmondsworth: Penguin, 1979.

Zilsel, Edgar. *Die Entstehung des Geniebegriffes. Ein Beitrag zur Ideengeschichte der Antike und des Frühkapitalismus.* Tübingen: J. C. B. Mohr (Paul Siebeck), 1926.

Zilsel, Edgar. *Die Geniereligion: Ein kritischer Versuch über das moderne Persönlichkeitsideal, mit einer historischen Begründung.* Vienna: Braumüller, 1918.

about the author

Alister McGrath is Andreas Idreos Professor of Science and Religion at Oxford University. After initial academic work in the natural sciences, McGrath turned to the study of theology and intellectual history, while occasionally becoming engaged in broader cultural debates about the rationality and relevance of the Christian faith. He has a long-standing interest in educational issues and has developed a series of theological textbooks that are widely used throughout the world.

Dr. McGrath is a bestselling author of more than fifty books and a popular speaker, traveling the world every year to speak at various conferences.